Glutamine and Glutamate in Mammals

Volume II

Editor

Elling Kvamme, M.D., Dr. med.

Professor of Neurochemistry
Neurochemical Laboratory
University of Oslo
Oslo, Norway

CRC Press
Taylor & Francis Group
Boca Raton London New York

CRC Press is an imprint of the
Taylor & Francis Group, an **informa** business

First published 1988 by CRC Press
Taylor & Francis Group
6000 Broken Sound Parkway NW, Suite 300
Boca Raton, FL 33487-2742

Reissued 2018 by CRC Press

Library of Congress Cataloging-in-Publication Data

Glutamine and glutamate in mammals.

Includes bibliographies and indexes.
1. Glutamine--Metabolism. 2. Glutamic acid--Metabolism. 3. Mammals--Physiology. I. Kvamme, Elling, 1918-
QP562.G55G57 1988 599'.019'245 87-21788
ISBN 0-8493-6856-1 (v. 1)
ISBN 0-8493-6857-X (v. 2)

A Library of Congress record exists under LC control number: 87021788

Publisher's Note
The publisher has gone to great lengths to ensure the quality of this reprint but points out that some imperfections in the original copies may be apparent.

Disclaimer
The publisher has made every effort to trace copyright holders and welcomes correspondence from those they have been unable to contact.

ISBN 13: 978-1-315-89318-1 (hbk)
ISBN 13: 978-1-351-07228-1 (ebk)

Visit the Taylor & Francis Web site at http://www.taylorandfrancis.com and the
CRC Press Web site at http://www.crcpress.com

PREFACE

The present comprehensive volume, which is based on the joint effort of a great many top scientists and covering most aspects of the metabolism and function of glutamine and glutamate in mammals, is a result of what we felt was a specific demand. Thus, the majority of previous books and review articles on glutamine and glutamate are confined to limited areas (e.g., energy metabolism, nervous transmission, etc.) or organs (e.g., brain, kidney, liver, etc.). This is not surprising in view of the vast amount of recent publications in each field, but it contributes to some sort of compartmentalization — to use a popular word — that creates barriers and subspecialization. The present volumes intend to act as a carrier between the various compartments. Similar to biological carriers, this one must be selective, since every discovery claimed to be done cannot be and does not deserve to be conveyed. However, what should be included will always be a matter of different opinion. Anyway, to communicate essential information from the many interrelated subfields appears to be necessary, since quite a few recent publications reflect the lack of knowledge of relevant findings in neighboring fields.

In addition to breaking down artificial barriers between subjects and fields, this volume also serves another purpose, namely to transmit different views on controversial matters. For that reason each author has been allowed to speak with his own voice and present his personal opinion on problems under debate. The careful reader will therefore discover that in spite of considerable overlapping in problems to be discussed, the flavor of the discussions may be different. Furthermore, although the reader may feel that he knows what is worth knowing about a subject after having read the relevant chapter, he may change his mind following the study of other chapters touching upon the same subject. However, the present volumes do not pretend to give the final answer to the many problems presented, but it intends to give a cross-section of a process under continuous development, and I wish to convey my sincere thanks to the authors who have all contributed to approach this goal.

Elling Kvamme

THE EDITOR

Dr. Elling Kvamme is professor of neurochemistry at the Oslo University and Head of the Neurochemical Laboratory.

In 1947, Dr. Kvamme received an M.D. at the Oslo University. Thereafter, he studied organic and physical chemistry, and in 1959 he received the scientific degree, Dr. med., at the Oslo University. After 3 years of service at Dikemark Hospital, Asker, and a 1-year internship (internal medicine and surgery) at the Ullevål Hospital, Oslo, he received a Fulbright/Smith Mundt Fellowship and worked as a Research Fellow at the Sloan-Kettering Institute for Cancer Research in New York from 1952 to 1954, and at the Public Health Research Institute of the City of New York, Department of Biochemistry, from 1954 to 1955. From 1955 to 1962, Dr. Kvamme was appointed Assistant Head of the Central Laboratory, Ullevål Hospital. Thereafter, he was appointed Head of the newly formed Neurochemical Laboratory at the Oslo University Psychiatric Clinic and, in 1966, appointed professor of Neurochemistry at the Oslo University. He is currently teaching neurochemistry to medical students and organizing postgraduate courses.

In 1962, Dr. Kvamme spent 6 months as a Technical Assistant Expert at the National Institute of Cancer, Rio de Janeiro, being appointed by the International Atomic Energy Agency, Vienna.

Dr. Kvamme has presented numerous papers at international meetings, as well as guest lectures at various universities and institutes in Europe and the U.S. He has also taken an active part in organizing many national and international meetings. Dr. Kvamme has served as the President of the Norwegian Biochemical Society from 1976 to 1978, the Treasurer of the International Society for Neurochemistry (ISN) from 1977 to 1981, the Chairman of ISN from 1981 to 1983, and the Chairman of the Policy Advisory Committee of ISN from 1985 to 1987.

Dr. Kvamme's main interest includes the metabolism and function of amino acids, in particular with regard to glutamine and glutamate. He has published a great number of scientific articles in international journals, in addition to several review articles in handbooks and scientific journals.

CONTRIBUTORS

M. H. Aprison
Distinguished Professor of Neurobiology and Biochemistry
Departments of Psychiatry and Biochemistry
Institute of Psychiatric Research
Indiana University School of Medicine
Indianapolis, Indiana

Soll Berl
Research Professor
Department of Neurology
Mt. Sinai School of Medicine
New York, New York

Henry F. Bradford M.R.C.Path.
Professor of Neurochemistry
Department of Biochemistry
Imperial College
London, England

Oliver W. Cass
Assistant Professor
Department of Medicine
Hennepin County Medical Center
Minneapolis, Minnesota

Jørgen Drejer
Department Head
Department of Cellular Neurochemistry
Ferrosan Research Division
Soeberg, Denmark

Frode Fonnum
Professor
Department of Environmental Toxicology
Norwegian Defense Research Establishment
Kjeller, Norway

Leif Hertz
Professor
Department of Pharmacology
College of Medicine
University of Saskatchewan
Saskatoon, Saskatchewan, Canada

Tage Honoré
Director
Neurochemica Laboratories
Ferrosan Research Division
Soeberg, Denmark

Thomas L. Perry
Professor
Department of Pharmacology and Therapeutics
University of British Columbia
Vancouver, British Columbia, Canada

Dale W. Peterson
Senior Lecturer
Division of Biological Sciences
Hatfield Polytechnic
Hatfield, England

Andreas Plaitakis
Associate Professor
Department of Neurology
Mount Sinai School of Medicine
New York, New York

Arne Schousboe
Professor
Department of Biochemistry A
Panum Institute
University of Copenhagen
Copenhagen, Denmark

Richard P. Shank
Research Fellow
Department of Biological Research
McNeil Pharmaceutical
Philadelphia, Pennsylvania

Mary J. Voaden
Reader in Biochemistry
Department of Visual Science
Institute of Ophthalmology
London, England

Leslie Zieve
Professor
Department of Medicine
Hennepin County Medical Center
Minneapolis, Minnesota

TABLE OF CONTENTS

Volume I

Volume II

GLUTAMINE AND GLUTAMATE IN THE CENTRAL NERVOUS SYSTEM

PATHOLOGY OF GLUTAMINE AND GLUTAMATE

Glutamine and Glutamate in the Central Nervous System

Chapter 15

GLUTAMATE AS A NEUROTRANSMITTER

Richard P. Shank and M. H. Aprison

TABLE OF CONTENTS

I. HISTORICAL OVERVIEW

The realization in the early 1950s that synaptic transmission in the CNS of vertebrates is predominantly chemically mediated provided impetus to discover the chemical agents mediating this function. One of the first systematic attempts to find substances that might serve such a function was undertaken by Curtis and colleagues.[1-3] Their studies revealed, among other things, that glutamate potently activates neurons. In their studies, glutamate excited virtually all neurons they tested. This ''universal'' excitatory action coupled with the ubiquitous occurrence of glutamate throughout the CNS, and in all tissues, was contrary to the prevailing view of what the biological characteristics of a neurotransmitter should be. Curtis and colleagues therefore tentatively discounted glutamate as a neurotransmitter.[4-7]

Meanwhile, parallel studies with invertebrates revealed that glutamate potently excites muscle contraction in arthropods.[8,9] Here also there was reluctance initially to accept glutamate as a likely neurotransmitter agent. However, studies by the Takeuchis and a host of other investigators[10-12] provided a solid foundation for the candidacy of glutamate as a neuromuscular excitatory transmitter in arthropods by the mid 1970s.

Interest in a possible neurotransmitter function of glutamate as well as aspartate and glycine in vertebrates was kindled by the studies of Aprison et al.[13-14] who in the mid 1960s undertook a systematic study to identify neurotransmitters in the spinal cord. Their approach, which was a combined biochemical and electrophysiological approach, demonstrated that the distribution of glutamate in the spinal cord and roots was consistent with a role for this amino acid as the neurotransmitter released by primary afferent neurons, i.e., glutamate is concentrated in primary afferent neurons and their collaterals in the dorsal gray matter.[15-17] Subsequent distribution studies by Johnson and Aprison[18] provided a neurochemical basis for regarding glutamate as a possible neurotransmitter in other areas of the vertebrate CNS. By the early 1970s, interest in glutamate as a possible neurotransmitter became widespread among neurochemists, neurophysiologists, neuropharmacologists, and other types of neurobiologists. Research pertaining to the neurotransmitter function of glutamate has continued to accelerate during the past 10 years. This has resulted in considerable progress in our understanding of the mechanistic nature of the neuroexcitatory action of glutamate and in the identification of neuronal pathways in which it may function as a neurotransmitter.

II. PHYSIOLOGICAL SIGNIFICANCE OF THE NEUROEXCITATORY ACTION OF GLUTAMATE

A recurrent issue has been whether or not glutamate is a true neurotransmitter or some type of neuromodulatory agent. This controversy has been fueled in part by the difficulty in rigorously establishing that glutamate meets all the accepted criteria for the neurotransmitter function. The increasing realization that the physiological effects and functional roles of many neuroactive agents may be complex and the ever-mounting evidence that some neurons may release two or more neuroactive agents have made it apparent that these substances cannot be neatly grouped into precise functional categories.[19-21] Hence, there is a growing tendency to refer to all endogenous neuroactive substances, whose action is likely to be receptor mediated, as neurotransmitters. The precise terminology that can properly classify neuroactive agents released by neurons may await a more complete understanding of the chemical events that underlie synaptic communication between neurons.

The numerous published studies dealing with the putative neurotransmitter function of glutamate have provided a wealth of confusing and seemingly conflicting data. Although some confusion may be attributed to the use of techniques that provide ambiguous or irrelevant data, a major source of confusion may be due to the reasonable tendency of neuroscientists to underestimate the complexity of biochemical and physiological activities in neural tissues. One aim of this review is to offer some plausible explanations for some of the confusion surrounding the functional significance of the neuroexcitatory action of glutamate.

III. THE NATURE OF THE NEUROEXCITATORY EFFECTS OF GLUTAMATE

Glutamate excites most neurons throughout the CNS of vertebrates,[21-26] but has little or no effect on peripheral autonomic or enteric neurons.[26] When appropriately administered, the onset of action and termination of response are both very fast. There is no doubt that glutamate interacts with specific receptors in the cell membrane, of which there are several types. Each type of receptor may differ in physiological as well as pharmacological respects.[27] Some types of receptors may be confined to, or at least concentrated in, the specialized postsynaptic membrane. One or more types may be restricted to the membrane of presynaptic terminals, and some types may reside in nonsynaptic regions.[28]

A. Glutamate Activates Membrane Depolarizing Channels

Numerous studies have demonstrated that the excitatory action of glutamate is due to a rapid membrane depolarization induced by an influx of Na^+ and in some instances Ca^{2+}.[25,29-34] Thus, it can be inferred that postsynaptic excitatory glutamate receptors are functionally linked to ion channels selective for certain cations. Current knowledge indicates that there are at least two major types of channels, one of which may selectively pass Na^+ and K^+, whereas the other may also pass Ca^{2+}.[32-34] At some synapses the latter channel is linked to *N*-methyl-D-Aspartate (NMDA) receptors (NMDA activates these receptors selectively).[32] This channel can be blocked by Mg^{2+}, which may function as a physiological modulator.[32-35] Some available evidence suggests that the two known types of channels are not necessarily linked to a specific pharmacological type of receptor.[34,35]

B. Glutamate and the Synthesis of Second Messengers

Glutamate, at comparatively high concentrations (~1 m*M*), induces an increase in the level of cyclic GMP (cGMP) in brain slice preparations.[36-38] This is consistent with the concept that cGMP is a second messenger at some glutamate receptors. The glutamate autoreceptor, whose existence is controversial,[27] would be a likely candidate. Another possibility is that the increase in cGMP occurs secondarily to the membrane depolarization process. Certainly the influx of Na^+ and Ca^{2+} and the associated depolarization would be expected to activate a variety of enzyme-mediated events.

Glutamate can also enhance the hydrolysis of inositol phospholipids, thereby generating inositol phosphates.[39,40] However, as with cGMP formation, the possibility exists that this biochemical effect is secondary to Na^+ and Ca^{2+} influx. In this respect it is noteworthy that K^+-induced depolarization can increase inositol phosphate formation.[41]

C. Comparison of Glutamate-Induced Excitation to Neurally Evoked Excitation

A number of studies have been performed on specified neurons to determine if the postsynaptic response elicited by glutamate is identical to that elicited by the evoked release of the endogenous neurotransmitter. In most studies undertaken to test this "identity of action", the technique used has been to establish the precise membrane potential at which current flowing through the activated postsynaptic channels has a net value of zero (the reversal potential or sometimes referred to as the equilibrium potential). Where this identity of action criterion has been tested, the reversal potential for glutamate was usually somewhat more negative than that for the endogenous agent.[25,26,29,31] However, it must be emphasized that for a variety of reasons the results of these experiments using intact CNS tissues are not definitive. Almost invariably, iontophoretically applied glutamate impinges on a somewhat different area of the cell surface than the endogenously released transmitter, thereby potentially activating a different population of receptors. Frequently, glutamate is applied near the cell soma, whereas the neurally activated excitatory synapses may be predominantly or exclusively on dendrites. Since the membrane potential and current flowing through the membrane are measured with an intracellular electrode inserted in or near the cell soma, the recorded reversal potential for synaptic current at distal regions of dendrites may be biased due to a loss of current flowing between the synapse and the electrode (the reversal potential will be biased positive). A further complication is the voltage-dependent rectifying properties of neuronal membranes, which tend to obscure the increase in membrane conductance.[42]

Analysis of membrane "noise" and the use of "patch clamp" techniques, which yield information pertaining to channel activation and inactivation characteristics, can provide more definitive information regarding the identity of action.[43] Although these techniques have been applied successfully to invertebrate preparations and neurons in culture, these techniques are difficult to use in intact CNS tissues.

Some studies in which the activity of glutamate applied by bulk flow was compared to neurally evoked excitatory postsynaptic responses have yielded some rather puzzling results. In one study, an isolated chick retina-optic nerve preparation was used,[44] and, in another, rat brain hippocampal slices with an intact perforant path were used.[45] In both studies, glutamate was applied at sufficiently high steady-state concentrations (0.1 to 1 m*M*) that the glutamate response eventually faded away; however, neurally evoked excitatory responses could still be elicited. Although the authors reasonably concluded that the data argued against a transmitter function for glutamate at those synapses, other reasonable conclusions are possible. For example, glutamate receptors may possess the ability to accommodate in a way that allows them to shift to a very low affinity for glutamate. Thus, during periods of exceedingly high extracellular glutamate levels, the receptors could still respond to synaptically released glutamate, assuming the extracellular concentration during synaptic release exceeded the steady-state level. An alternative explanation is that a cotransmitter (e.g., aspartate) could activate the same or different receptors in the presence of high glutamate levels.

Another means of evaluating the identity of action criterion is to compare the effects of specific antagonists of glutamate on postsynaptic responses evoked by iontophoretically applied glutamate and responses evoked by neural stimulation. Moderately potent antagonists selective for specific types of receptors activated by glutamate have been developed only recently.[46,47] Despite this, the weak antagonist glutamate diethylester (GDEE) was found to inhibit both the action of glutamate and the transmitter released by cerebral cortical pyramidal neurons synapsing on neurons in the striatum.[48,49] In contrast, the excitatory effect of glutamate on lateral geniculate neurons was pharmacologically distinct from the action of the transmitter released from fibers in the optic tract;[50] similarly, the effect of glutamate on neurons in the olfactory cortex was different from that of the transmitter released by fibers in the lateral olfactory tract (Section VI).[51] However, because at least three pharmacological types of receptors exist that are activated by glutamate, these results must be interpreted cautiously. Iontophoretically applied glutamate may not have access to receptors activated by the endogenous transmitter as readily as it may to another population of a different pharmacological type. Consequently, glutamate could elicit a marked excitatory response without activating the specific receptors activated by the endogenously released neurotransmitter agent.

D. Glutamate Concentration and Neuronal Excitability

Although it would be of considerable value to know the precise concentration at which glutamate activates receptors at specific types of synapses, this is not yet known definitively.

When superfused onto intact tissues or cells in culture, the threshold concentration at which glutamate excites neurons is approximately 10^{-6} *M*, and the concentration required for maximal excitation is 10^{-4} to 10^{-3} *M*.[26,52,53] However, in intact CNS tissues, glutamate is rapidly removed from the extracellular fluid by transport systems in the membrane of synaptic terminals and astrocytes.[53-58] Therefore, the actual concentration at the receptors may be less than that in the bathing fluid.

It seems reasonable to assume that the steady-state extracellular concentration in resting state conditions is below the threshold for receptor activation; however, this concentration is not known precisely. Reported values for glutamate in cerebrospinal fluid (CSF) typically range between 1 and 10 μ*M*,[59-62] but it is possible that the extracellular concentration within the CNS parenchyma is less than this. The isolated toad brain, which is metabolically highly viable when properly treated,[63] will remove glutamate from a reservoir of artificial CSF until a steady-state concentration of 1 μ*M* is achieved.[64] Furthermore, if glutamate is omitted from the solution, a release of glutamate occurs until the same extracellular concentration is achieved.[65] The use of indwelling cannulas implanted in discrete CNS regions has provided data indicating that the steady-state extracellular concentration is about 0.3 μ*M*;[66] however, this may be low due to possible uptake by glia, which proliferate around the cannula. It seems reasonable to infer from these data that the normal steady-state concentration of glutamate is just below or at the threshold for receptor activation.

The maximum extracellular concentration achieved during synaptic activation is unknown and currently can only be conjectured. Considering the fact that the average concentration of glutamate in CNS tissues is nearly 10 m*M*, the concentration within synaptic terminals that release glutamate must be at least 10 m*M*, and within vesicles that accumulate glutamate the concentration might easily be 30 to 40 m*M*.[67,68] Thus, a reasonable estimate is that the maximum concentration achieved in the synaptic cleft under normal physiological conditions is between 0.1 and 1 m*M*.

E. Depolarizing Action of Glutamate on Glia Cells

Although early studies indicated that glia cells were not responsive to glutamate,[69] several recent studies have demonstrated that glutamate either directly or indirectly can depolarize some glia cells. Hosli and colleagues[70] concluded from studies with primary tissue cultures (explants) from the mouse spinal cord that the depolarizing action of glutamate on astrocytes in their cultures was due to an extracellular accumulation of K^+, itself due to the depolarizing effect of glutamate on neurons. However, more recent studies using either primary cultures derived from the rat brain or intact optic nerve from the *Necturus maculosa* indicate that glutamate can depolarize glia membranes directly.[71-73] In some instances this appears to be mediated via kainate receptors linked to channels that selectively pass Na^+.[71]

The functional significance of the depolarizing action of glutamate on glial cells is not known, but one possibility is that glutamate may evoke the release of taurine, which may cause a generalized inhibitory effect on nearby neurons.[74]

IV. BIOCHEMICAL PROCESSES THAT UNDERLIE THE NEUROTRANSMITTER FUNCTION

The metabolism of glutamate in CNS tissues is quite complex. This is due in part to the roles of glutamate in oxidative metabolism and nitrogen metabolism. The elucidation of the metabolism of glutamate as it relates specifically to the neurotransmitter function is a formidable challenge.

A. Metabolic Precursors of the Neurotransmitter Pool of Glutamate

Metabolites can serve as a precursor of the transmitter pool of glutamate if they can be a net source of either the glutarate or amino moieties. Due to the metabolic complexity of glutamate, at least ten potential precursors exist,[75] some of which can supply both the glutarate and amino moieties. Criteria for evaluating the role of a compound as a metabolic precursor include

1. The availability of the compound to glutamatergic terminals
2. The ability of glutamatergic terminals to accumulate (take up) the compound and metabolize it to glutamate
3. The degree to which the uptake and metabolic conversion is subject to biochemical regulation
4. The effect that limiting the supply of the potential precursor has on the content of glutamate and the amount released during neuronal stimulation

In order to gain an appreciation of the quantitative significance of the role of a compound in replenishing the neurotransmitter pool of glutamate, the latter criterion must be tested experimentally using conditions relevant to the normal physiological state.

To date only three compounds have been studied to any significant extent as possible precursors. These include glutamine, glucose [or α-ketoglutarate (2-oxoglutarate)], and ornithine (or arginine).

1. Glutamine as a Precursor of Neurotransmitter Glutamate

Studies to determine if glutamine is a precursor of the neurotransmitter pool of glutamate were initiated in the early 1970s. These studies were motivated by the prevalence of glutamine in CNS tissues, its close metabolic relationship to glutamate, and the unusual metabolic compartmentation known to exist in CNS tissues.[76] Observations made during the 1970s established that glutamine is readily taken up into presumed glutamatergic terminals by saturable transport systems and that it can be rapidly and extensively metabolized to glutamate, particularly a pool that is released during membrane depolarization.[77,78] The results of these studies and other studies involving the uptake of glutamate by glia cells, in conjunction with conclusions drawn from previous metabolic compartmentation studies,[79] provided the basis for the formulation of the concept that there is a net flux of glutamate from neurons to astrocytes and a counter flux of glutamine in the reverse direction to replenish glutamate lost from the neurons (Figure 1).[80] This conceptual model, which is frequently referred to as "the glutamine cycle", has yet to be fully validated in terms of its quantitative significance.

In recent years, attention has been focused on the third and fourth criteria discussed above, i.e., the regulation of the conversion of glutamine to glutamate, and the coupling between the availability of glutamine to nerve terminals and the amount of glutamate released under physiologically relevant conditions. Of particular value are the studies of Nicklas[81] and Rothstein and Tabakoff,[82,83] which have provided compelling evidence that glutamine derived from astrocytes is a significant metabolic precursor of neurotransmitter glutamate in the cerebellum and striatum. A notable in vivo study by Thanki et al.[84] revealed that glutamine, when superfused onto the cortex of awake and unrestrained rats, was converted to glutamate that was selectively released by depolarizing stimuli.

2. Glucose and α-Ketoglutarate (2-Oxoglutarate) as Metabolic Precursors of Transmitter Glutamate

Interest in glucose as a precursor of the neurotransmitter pool of glutamate arose from the fact that a high percentage of glucose carbon passes through glutamate during normal oxidative metabolism in CNS tissues.[76] However, an important consideration that was ignored in many studies was the requirement that glucose be metabolized via an anaplerotic pathway in order to serve as a net source of the dicarboxylate (glutarate) moiety of glutamate.

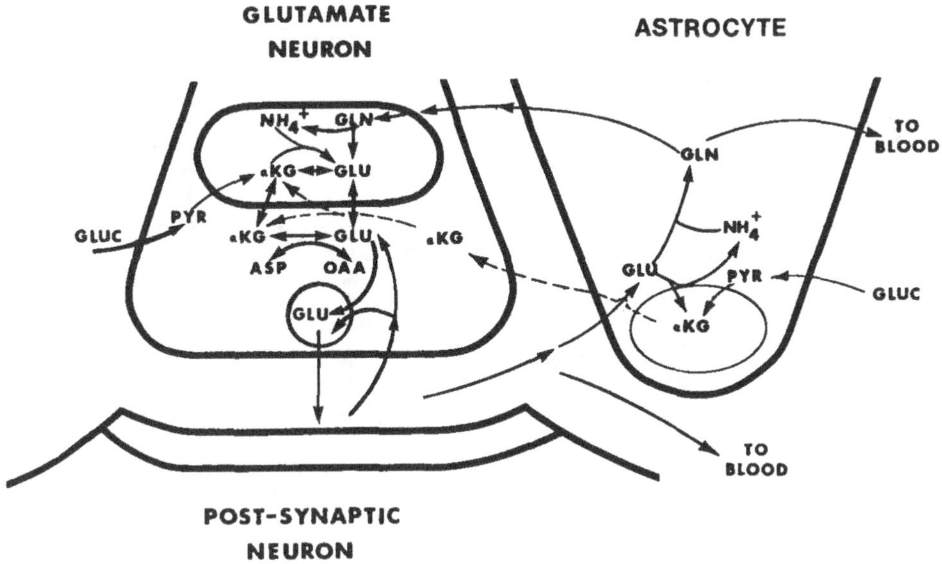

FIGURE 1. A model depicting some of the biochemical events thought to be associated with the neurotransmitter function of glutamate. This model assumes that glutamate is accumulated into synaptic vesicles from which it is released during synaptic activation. The relative proportion of the released glutamate that is returned to the synaptic terminal in comparison to the amount taken up by astrocytes or lost to the blood is not known.

Metabolic studies indicated that anaplerotic activity in the form of CO_2 fixation and pyruvate carboyxlation occurred predominantly or exclusively in the small metabolic compartment elucidated by Berl and colleagues.[76,85] The subsequent demonstration that glutamine synthetase, a marker for the small metabolic compartment, is an astrocyte-specific enzyme in CNS tissues[86] provided compelling evidence that anaplerotic activity occurs primarily in astrocytes and not in synaptic terminals. This has been confirmed by the demonstration that pyruvate carboxylase, the predominant anaplerotic enzyme in CNS tissues,[87] is an astrocyte-specific enzyme.[88,89]

The evidence that anaplerotic activity is restricted to astrocytes prompted us to hypothesize that astrocytes serve as a source of one or more citrate cycle intermediates for glutamatergic synaptic terminals (Figure 2). This hypothesis predicts that astrocytes must generate and release one or more citrate cycle intermediates and that synaptic terminals must avidly accumulate (net uptake) these intermediates. The latter prediction has been fulfilled in that synaptic terminals have been shown to avidly accumulate α-ketoglutarate and malate, but not citrate, by high-affinity transport systems.[90,91] Although a net release of α-ketoglutarate and malate from astrocytes has yet to be shown, most of the four criteria discussed earlier have been met for α-ketoglutarate. This metabolite is present in CSF at a concentration of 10 to 20 μM,[98] is avidly accumulated in net amounts by a Na^+-dependent, high-affinity system ($K_m \sim 1$ μM),[87] is rapidly metabolized to glutamate,[93] and when glutamine is present α-ketoglutarate can lead to a net increase in the content of glutamate in synaptic terminals.[93] Furthermore, the uptake of α-ketoglutarate is subject to biochemical regulation.[94] The quantitative contribution of α-ketoglutarate to the replenishment of the neurotransmitter pool glutamate remains to be established. In this respect, it is noteworthy that as much as 10% of the glucose metabolized in CNS tissues passes through the pyruvate carboxylase-mediated anaplerotic pathway.[95]

It must be emphasized that α-ketoglutarate can supply only the dicarboxylate moiety. Both glutamine and ornithine are ideal sources of the amino moiety required by α-ketoglutarate.

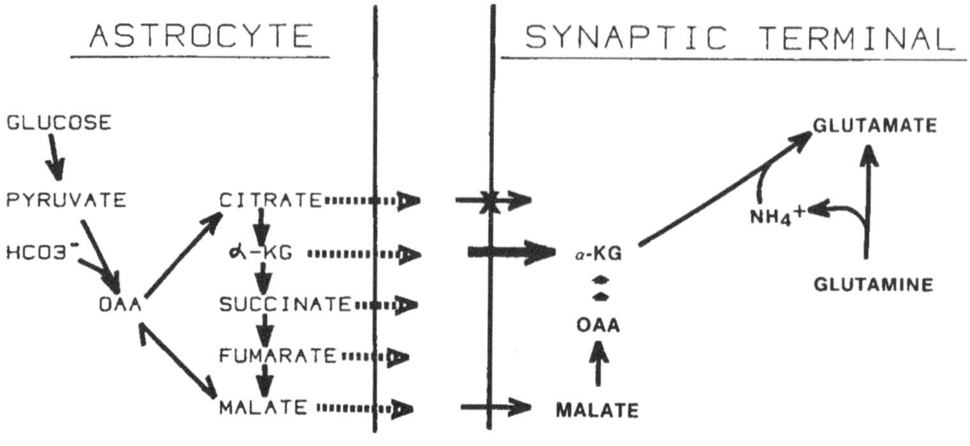

FIGURE 2. Possible significance of anaplerotic activity in astrocytes in the replenishment of the neurotransmitter pool of glutamate. Conceivably, any citrate cycle intermediate could be supplied to neurons; however, current evidence indicates that citrate is not taken up by synaptic terminals. No investigation of citrate cycle intermediates released from astrocytes has been reported. Note that glutamine (or another amino donor) must be available to supply the amino moiety.

3. Arginine and Ornithine as Precursors of Transmitter Glutamate

Several studies have established that these amino acids can be metabolized to glutamate by CNS tissues.[96-98] The presumed source of these potential precursors is the blood, although CNS tissues do have the capacity to form arginine from citrulline.[99] The apparent availability, rate of uptake by synaptosomes, and metabolic conversion to glutamate, when compared to corresponding data for glutamine and α-ketoglutarate, indicate that quantitatively the role of arginine and ornithine as precursors of glutamate may be of minor significance.[97] However, Roberts[100] has argued that the turnover rate of the transmitter pool may be slow in vivo, perhaps much slower than the turnover rate of the metabolic pools of glutamate.

A recent study of considerable interest demonstrated that L-canaline, a potent inhibitor of ornithine aminotransferase, when injected into the septum caused a gradual decrease of 20 to 40% in the concentration of glutamate during a 2-hr period.[101] This decrease was presumed to be due to the demonstrated inhibition of ornithine aminotransferase. However, this effect must be interpreted cautiously until possible effects of L-canaline on other biochemical systems involving glutamate are excluded.

4. Concluding Remarks on Metabolic Precursors of the Transmitter Pool of Glutamate

It should be emphasized that the replenishment of the neurotransmitter pool of glutamate is likely to be mediated through the utilization of several precursors, and the quantitative significance of each may vary depending on the metabolic condition of the tissue and the specific region within the CNS. Assuming that glutamate is stored in and released from vesicles (Section IV.B), the transmitter pool of glutamate must reside within vesicles. However, the ultimate origin of most of these molecules is likely to be mitochondrial since glutaminase, glutamate dehydrogenase, and ornithine aminotransferase are located in this organelle. It seems reasonable to assume that vesicles accumulate glutamate molecules from their immediate evnironment regardless of the origin of the molecule.

B. The Cellular Origin of the Transmitter Pool and the Release Mechanism

By the time glutamate became established as a putative neurotransmitter, the quantal nature of transmitter release was well documented, as was the cellular localization of monoamine neurotransmitters and acetylcholine in synaptic vesicles (or granules). Until recently, studies to determine if glutamate is enriched in synaptic vesicles were essentially negative.[102,103] Glutamate was present in isolated vesicles, but not in the amounts expected for a transmitter being secreted from these organelles. However, the recent demonstrations that isolated vesicles possess an ATP- and Cl^--dependent translocation (transport) process that can accumulate glutamate support the concept that glutamate is concentrated within vesicles in the living state.[67,68] The conditions used to isolate synaptic vesicles, perhaps the absence of Cl^-, may cause glutamate to "leak out" during the subcellular fractionation process.

Present knowledge is therefore consistent with the concept that glutamate is released quantally from vesicles during the excitation-secretion process. The number of quanta and amount of glutamate released by an action potential can only be conjectured; however, at one type of synapse where glutamate may be a transmitter (primary afferent neuronal synapses on motoneurons in the spinal cord), the number of quanta is small (approximately one to three).[104]

C. The Mechanism of Transmitter Inactivation

Subsequent to the demonstration that synaptic terminals possess Na^+-dependent, high-affinity membrane transport systems that rapidly remove glutamate from extracellular fluid, the concept was proposed that this uptake process serves as the mechanism for inactivating glutamate subsequent to its release.[54,55] The previous observations that glutamate, when applied to intact CNS tissues, is rapidly metabolized to glutamine,[76] coupled with the demonstration that glutamine synthetase is an astrocyte-specific enzyme,[86] suggest that astrocytes may be instrumental in removing glutamate from the extracellular fluid. This is supported by the observations that astrocyte membranes do posess Na^+-dependent, high-affinity glutamate transport carriers.[56-58,105]

Based on current values for the K_m and V_{max} for GABA uptake by synaptosomes, Cupello and Hyden[106] have calculated that uptake per se is not nearly fast enough to account for transmitter inactivation in its most precise sense, i.e., removal of molecules from the immediate vicinity of the receptors in the postsynaptic membrane. Based on current K_m and V_{max} data for glutamate, which are similar to those for GABA,[58,105] the same conclusion may be valid also for glutamate. A likely mechanism for the inactivation of glutamate is diffusion away from the postsynaptic receptors, made possible by the maintenance of very low extracellular glutamate levels by uptake systems present both in glutamatergic synaptic membranes and in astrocyte membranes.

V. TYPES OF NEURONS THAT UTILIZE GLUTAMATE AS A NEUROTRANSMITTER

Currently, there is no established marker that definitively identifies glutamatergic neurons. However, evidence that certain types of neurons are glutamatergic has been obtained using a variety of neurochemical and histochemical procedures. These include measurements of the content of glutamate in discrete samples of normal and lesioned tissue to demonstrate that glutamate is selectively concentrated in certain neurons and their terminals, membrane transport (uptake) experiments to demonstrate a selective accumulation of labeled substrate (e.g., L-glutamate or D-aspartate) by certain types of neurons, the release of endogenous or labeled exogenous glutamate (or D-aspartate) from CNS tissue *in situ*, tissue slices, or synaptosomes, and immunocytochemical localization of glutaminase, glutamate dehydrogenase, aspartate aminotransferase, and glutaraldehyde-fixed glutamate. The immunocytochemical localization of glutaraldehyde-fixed glutamate is a particularly promising technique developed originally by Storm-Mathisen and Ottersen[107] and recently used in modified forms by Madl and colleagues[108] and by Conti and colleagues.[109]

Current evidence indicates that several different types of neurons represented in nearly

Table 1
NEURONS IN WHICH GLUTAMATE MAY FUNCTION AS A TRANSMITTER OR A COTRANSMITTER

Types of neurons	Comments	Ref.[a]
Primary sensory systems		
Somatic sensory neurons	Glutamate is concentrated in spinal dorsal roots	13—17
	D-Aspartate is selectively accumulated by some dorsal root ganglion cell bodies	121
Afferents from baroreceptors	Lesions reduce glutamate uptake in the nucleus tractus solitarius	122
Auditory system: hair cells, primary afferents to cochlear nucleus, and granule cells in cochlear nucleus	Evidence includes selective accumulation of ^{3}H-D-aspartate and histochemical data	110—113
Olfactory bulb: mitral cells that project to the olfactory cortex	Glutamate does not mimic the endogenous transmitter pharmacologically	51, 119
Retina: some photoreceptor cells and some ganglion (optic tract) neurons	See Chapter 19	
Cerebral Cortex: projection neurons to subcortical regions and other regions	See Chapters 18 and 19 for detailed discussion	110—112, 114
Hippocampal formation: perforant afferents, commissural fibers, Schaffer collaterals, mossy fibers, and projection neurons to septum	See Chapter 18	
Cerebellum: granule cells (parallel fibers), climbing fiber afferents, and efferents from deep nuclei	See Chapters 18 and 19 for evidence that parallel and climbing fibers release glutamate	110—112, 115
	Immunocytochemical evidence for deep nuclei efferents is reported	108
Spinal cord and brainstem: some interneurons	See Chapters 18 and 19 for a detailed discussion	16, 110—113

[a] References 110 to 112 and Chapters 18 and 19 are review articles; other references present observations not discussed previously in review articles.

every major region in the CNS are glutamatergic (Table 1). Most of these are projection neurons, but some excitatory interneurons appear to be glutamatergic also. Several recent reviews, including Chapters 18 and 19, provide detailed discussions of the evidence supporting glutamate as the transmitter released by specific types of neurons.[110-112]

VI. TYPES OF NEURONS THAT MAY UTILIZE GLUTAMATE AS A COTRANSMITTER

It is becoming increasingly evident that at many, if not most, synapses the transmitted signals involve more than just a rapid on-off switching process. Instead, a complex pattern of postsynaptic membrane responses may be elicited by the activation of single types of synapses.[19,20,116] Relevant to this is the rapidly accumulating evidence that many types of neurons release more than one neuroactive substance, one of which is nearly always a peptide.[19-21] Although little information is presently available regarding the colocalization of glutamate with peptides, some evidence suggests that at least two types of neurons may use glutamate as a cotransmitter. These are the mitral cells that project from the olfactory bulb to the olfactory cortex and the small unmyelinated primary afferent neurons (the C fibers). The mitral cells contain high amounts of glutamate and aspartate, avidly accumulate both amino acids, and release both.[117] These cells may also contain the dipeptide *N*-acetylaspartyl-glutamate (NAAG).[118] When iontophoretically applied onto neurons to which the mitral cells project, both amino acids and NAAG elicited action potentials; however, only the excitatory responses elicited by NAAG were affected pharmacologically in a manner similar to the response elicited by electrical stimulation of the lateral olfactory tract.[119] These results are not consistent with the view that glutamate functions as the transmitter at these synapses. However, these observations must be interpreted cautiously, due in part to the possible experimental complications noted in Section III.C, but also because the natural response (the EPSP) might be due to the combined action of glutamate and aspartate or either amino acid and NAAG or a combination of all three. With regard to this possibility, the contribution of any one of the three may vary depending on the frequency of firing and temporal pattern of mitral cell activity and on the metabolic state of these cells. As discussed later in Section VII, there is compelling evidence that transmission at excitatory neuromuscular junctions in the walking limbs of lobsters is mediated by glutamate and aspartate.

The higher concentration of glutamate in the dorsal root as compared to the ventral root in cats and rats,[13-16,120] the reduced content of glutamate in the dorsal horn of the lumbar region of the spinal cord after the dorsal root is surgically severed,[16] and the selective accumulation of D-aspartate into some dorsal root ganglion cells[121] comprise some of the evidence that glutamate is the transmitter released by primary afferent neurons. Substance P is also concentrated in primary afferent neurons and is known to be localized primarily to the small nonmyelinated C fibers.[123] Glutamate and substance P may be colocalized in some C fibers,[121] and both excite neurons onto which the C fibers terminate. A characteristic of these neurons is that they generate both fast and slow EPSPs in response to stimulation of primary afferent neurons, and the slow EPSPs selectively disappear after capsaicin induced lesions.[124] A postulate consistent with these results is that the fast EPSPs are mediated by glutamate and the slow EPSPs are mediated by substance P or a combination of glutamate and substance P. In this respect, it is of interest that substance P has little direct excitatory activity on rat spinal motoneurons, but can markedly potentiate and prolong the excitatory action of glutamate.[125]

Regarding the possible role of glutamate as a cotransmitter at some synapses, it is noteworthy that thyrotropin-releasing horomone (TRH) can either enhance or inhibit the action of glutamate on spinal motoneurons,[126] and arginine-vasopression can enhance the activity of glutamate on some neurons in the lateral septum.[127]

VII. GLUTAMATE AS A NEUROTRANSMITTER IN INVERTEBRATES

There is virtually no doubt that glutamate functions as either a transmitter or a cotransmitter at some types of excitatory neuromuscular junctions in a variety of arthropods.[127,128] Glutamate may also be a neurotransmitter at some synapses in mollusks such as the squid and *Aplysia*[127-129] and even in species as primitive as the sea anemone.[130]

At excitatory neuromuscular junctions in the walking limbs of the lobster *Homarus americanus*, synaptic transmission is probably mediated by glutamate and aspartate functioning synergistically. Aspartate is weakly excitatory by itself, but when applied together with glutamate, the excitatory activity of glutamate can be markedly potentiated by aspartate at concentrations that would otherwise be inactive.[131] Aspartate is present in the excitatory axons at concentrations three times that of glutamate and appears to be released from the nerve terminals during stimulation at a similar ratio.[132] Furthermore, D-aspartate selectively antagonizes the potentiative effect of L-aspartate, whereas D,L-4-amino-phosphonobutyrate selectively antagonizes the action of glutamate.[133] Both antagonists reduce the amplitude of neurally evoked EPSPs. L-aspartate has been shown to potentiate the excitatory activity of glutamate at some other invertebrate synapses, but the potentiation appears to be mechanistically different.[134,135]

VIII. CONCLUDING REMARKS

It is becoming increasingly evident that synaptic transmission at many synapses is a complex process involving more than one neuroactive agent. At some excitatory synapses, glutamate may function in concert with aspartate, NAAG, substance P, or another peptide to yield postsynaptic responses that vary in magnitude and duration depending on the frequency of firing of the presynaptic neuron and its temporal pattern of activity. Recent advances in the elucidation of biochemical events associated with the neurotransmitter release process provide a basis for suggesting that different neuroactive agents present within a given neuron can be stored in separate vesicles and released differentially as a function of time and frequency of synaptic activation.[136] Although the current practice of classifying neurons by the neurotransmitters they release is useful, this practice should not become a stumbling block to the pursuit of identifying other neuroactive agents serving a cotransmitter or neuromodulatory function. In this respect, glutamate may serve different functions at different synapses. If it turns out that neurons typically release more than one neuroactive compound, the "identity of action" criterion will be even more difficult to test since the precise nature of the postsynaptic response at a given synapse may vary depending on the relative contribution of each neuroactive agent.

ACKNOWLEDGMENT

We appreciate the able assitance of Ms. Maria Ciliberto in the preparation of this manuscript.

REFERENCES

1. **Curtis, D. R., Phillis, J. W., and Watkins, J. C.,** The depression of spinal neurons by γ-amino-n-butyric acid and β-alanine, *J. Physiol. (London)*, 146, 185, 1959.
2. **Curtis, D. R., Phillis, J. W., and Watkins, J. C.,** The chemical excitation of spinal neurons by certain acidic amino acids, *J. Physiol. (London)*, 150, 656, 1960.
3. **Curtis, D. R. and Watkins, J. C.,** The excitation and depression of spinal neurons by structurally related amino acids, *J. Neurochem.*, 6, 117, 1960.
4. **Curtis, D. R.,** The pharmacology of central and peripheral inhibition, *Pharmacol. Rev.*, 15, 333, 1963.
5. **Curtis, D. R. and Walkins, J. C.,** Acidic amino acids with strong excitatory actions on mammalian neurons, *J. Physiol. (London)*, 166, 1, 1963.
6. **Crawford, J. M. and Curtis, D. R.,** The excitation and depression of mammalian cortical neurons by amino acids, *Br. J. Pharmacol. Chemother.*, 23, 313, 1964.
7. **Curtis, D. R. and Watkins, J. C.,** The pharmacology of amino acids related to gamma-aminobutyric acid, *Pharmacol. Rev.*, 17, 347, 1965.
8. **Robbins, J.,** The excitation and inhibition of crustacean muscle by amino acids, *J. Physiol.*, 148, 39, 1959.
9. **van Harreveld, A.,** Physiological effects of glutamate and some related compounds, in *Inhibition in the Nervous System and Gamma-Aminobutyric Acid*, Roberts, E., Baxter, C. F., van Harreveld, A., Wiersma, C. A. G., Adey, W. R., and Killam, K. F., Eds., Pergamon Press, Elmsford, N.Y., 1960, 454.
10. **Takeuchi, A. and Takeuchi, N.,** The effect on crayfish muscle of iontophoretically applied glutamate, *J. Physiol. (London)*, 170, 296, 1964.
11. **Usherwood, P. N. R. and Machili, P.,** Pharmacological properties of excitatory neuromuscular synapses in the locust, *J. Exp. Biol.*, 49, 341, 1968.

12. **Kravitz, E. A., Slater, C. R., Takahashi, K., Bownds, M. D., and Grossfeld, R. M.,** Excitatory transmission in invertebrates — glutamate as a potential neuromuscular transmitter compound, in *Excitatory Synaptic Mechanisms,* Andersen, P. and Jansen, J. K. S., Eds., Universitetforlaget, Oslo, 1970, 85.
13. **Aprison, M. H., Graham, L. T., Jr., Livengood, D. R., and Werman, R.,** Distribution of glutamic acid in the cat spinal cord and roots, *Fed. Proc. Fed. Am. Soc. Exp. Biol.,* 24, 462, 1965.
14. **Aprison, M. H. and Werman, R.,** A combined neurochemical and neurophysiological approach to identification of central nervous system transmitters, in *Neurosciences Research,* Ehrenpreis, S. and Solnitzky, O. S., Eds., Academic Press, New York, 1968, 143.
15. **Graham, L. T., Jr., Shank, R. P., Werman, R., and Aprison, M. H.,** Distribution of some synaptic transmitter suspects in cat spinal cord: glutamic acid, aspartic acid, γ-amino-butyric acid, glycine, and glutamine, *J. Neurochem.,* 14, 465, 1967.
16. **Davidoff, R. A., Graham, L. T., Jr., Shank, R. P., Werman, R., and Aprison, M. H.,** Changes in amino acid concentrations associated with loss of spinal interneurons, *J. Neurochem.,* 14, 1025, 1967.
17. **Johnson, J. and Aprison, M. H.,** The distribution of glutamic acid, a transmitter candidate, and other amino acids in the dorsal sensory neuron of the cat, *Brain Res.,* 24, 285, 1970.
18. **Johnson, J. and Aprison, M. H.,** The distribution of glutamate and total free amino acids in thirteen specific regions of the cat central nervous system, *Brain Res.,* 26, 141, 1971.
19. **Lundberg, J. M. and Hokfelt, T.,** Coexistence of peptides and classical neurotransmitters, *Trends Neurosci.,* 6, 325, 1983.
20. **O'Donohue, T. L., Millington, W. R., Handelman, G. E., Contreras, P. C., and Chronwall, B. M.,** On the 50th Anniversary of Dale's law: multiple neurotransmitter neurons, *Trends Pharmacol. Sci.,* 6, 304, 1985.
21. **Wise, S. P.,** Co-localization of GABA and neuropeptides in the cerebral cortex, *Trends Neurosci.,* 8, 92, 1985.
22. **Krnjevic, K. and Phillis, J. W.,** Iontophoretic studies of neurones in the mammalian cerebral cortex, *J. Physiol. (London),* 165, 274, 1963.
23. **Curtis, D. R., Duggan, A. W., Felix, D., Johnston, G. A. R., Tebecis, A. K., and Watkins, J. C.,** Excitation of mammalian neurones by acidic amino acids, *Brain Res.,* 41, 283, 1972.
24. **Curtis, D. R. and Johnston, G. A. R.,** Amino acid transmitters in the mammalian central nervous system, *Ergeb. Physiol., Biol. Chem. Exp. Pharmakol.,* 69, 188, 1974.
25. **Krnjevic, K.,** Chemical nature of synaptic transmission in vertebrates, *Physiol. Rev.,* 54, 418, 1974.
26. **Johnson, J. L.,** The excitant amino acids glutamic and aspartic acid as transmitter candidates in the vertebrate central nervous system, *Prog. Neurobiol.,* 10, 155, 1978.
27. **Roberts, P. J.,** Presynaptic receptors for excitatory amino acid transmitters, in Excitatory Amino Acids, Roberts, P. J., Storm-Mathisen, J., and Bradford, H. F., Eds., MacMillan, London, 1986, 197.
28. **Freeman, A. R.,** Polyfunctional role of glutamic acid in excitatory synaptic transmission, *Prog. Neurobiol.,* 6, 137, 1976.
29. **Curtis, D. R.,** The actions of amino acids upon mammalian neurones, in *Studies in Physiology,* Curtis, D. R. and McIntyre, A. K., Eds., Springer-Verlag, New York, 1965, 34.
30. **Barker, J. L. and Nicoll, R. A.,** The pharmacology and ionic dependency of amino acid responses in the frog spinal cord, *J. Physiol. (London),* 228, 259, 1973.
31. **Puil, E.,** S-glutamate: its interactions with spinal neurons, *Brain Res. Rev.,* 228, 229, 1981.
32. **MacDonald, J. F., Proietis, A. V., and Wojtowicz, J. M.,** L-Aspartic acid induces a region of negative slope conductance in the current voltage relationship of cultured spinal cord neurones, *Brain Res.,* 237, 248, 1982.
33. **Mayer, M. L. and Westbrook, G. L.,** The action of N-methyl-D-aspartic acid on mouse spinal neurones in culture, *J. Physiol. (London),* 361, 65, 1985.
34. **Yamamoto, C. and Sato, H.,** Some properties of ionic channels activated by excitatory amino acids in hippocampal neurons, *Exp. Brain Res.,* 57, 313, 1985.
35. **Nistri, A., Arenson, M. S., and King, A.,** Excitatory amino acid-induced responses of frog motoneurones bathed in low Na^+ media: an intracellular study, *Neuroscience,* 14, 921, 1985.
36. **Ferrendelli, J. A., Chang, M. M., and Kinscherf, D. A.,** Elevation of cyclic GMP levels in central nervous system by excitatory and inhibitory amino acids, *J. Neurochem.,* 22, 535, 1974.
37. **Garthwaite, J.,** Excitatory amino acid receptors and guanosine 3′,5′ cyclic monophosphate in incubated slices of immature and adult rat cerebellum, *Neuroscience,* 7, 2491, 1982.
38. **Foster, G. A. and Roberts, P. J.,** Excitatory amino acid receptor stimulation: involvement of cyclic GMP?, in *CNS Receptors — From Molecular Pharmacology to Behavior,* Mandel, P. and DeFeudis, F. V., Eds., Raven Press, New York, 1983, 175.
39. **Sladeczek, F., Jean-Philippee, P., Recasens, M., Bockaert, J., and Weiss, S.,** Glutamate stimulates inositol phosphate formation in striatal neurones, *Nature (London),* 317, 717, 1985.

40. **Nicoletti, F., Meek, J. L., Iadarola, M. J., Chuang, D. M., Roth, B. L., and Costa, E.,** Coupling of inositol phospholipid metabolism with excitatory amino acid recognition sites in rat hippocampus, *J. Neurochem.*, 46, 40, 1986.
41. **Kendal, D. A. and Nakorski, S. R.,** Inositol phospholipid hydrolysis in rat cerebral cortical slices. II. Calcium requirement, *J. Neurochem.*, 42, 1388, 1984.
42. **Nelson, P. G. and Frank, K.,** Anomalous rectification in cat spinal motoneurons and effect of polarizing currents on excitatory postsynaptic potential, *J. Neurophysiol.*, 30, 1097, 1967.
43. **Anderson, C. R., Cull-Candy, S. G., and Miledi, R.,** Glutamate current noise: postsynaptic channel kinetics investigated under voltage clamp, *J. Physiol. (London)*, 228, 219, 1978.
44. **Trubatch, J., Verhulst, F. C., and van Harreveld, A.,** Glutamate as a transmitter: comparison between the crustacean neuromuscular junction and the chicken retina, *Comp. Biochem. Physiol. A*, 45, 183, 1973.
45. **Tagni, L., Baudry, M., and Lynch, G.,** Desensitization to glutamate does not affect synaptic transmission in rat hippocampal slices, *Brain Res.*, 261, 167, 1983.
46. **Watkins, J. C.,** Excitatory amino acids and central synaptic transmission, *Trends Pharmacol. Sci.*, 5, 373, 1984.
47. **Fagg, G. E.,** L-Glutamate, excitatory amino acid receptors and brain function, *Trends Neurosci.*, 6, 207, 1985.
48. **Spencer, H. J.,** Antagonism of cortical excitation of striatal neurons by glutamic acid diethylester: evidence for glutamic acid as an excitatory transmitter in the rat striatum, *Brain Res.*, 102, 91, 1976.
49. **Stone, T. W.,** Blockade by amino acid antagonists of neuronal excitation mediated by the pyramidal tract, *J. Physiol. (London)*, 257, 187, 1976.
50. **Tebecis, A. K.,** Studies on the identity of the optic nerve transmitter, *Brain Res.*, 63, 31, 1973.
51. **Hori, N., Auter, C. R., Braitman, D. J., and Carpenter, D. O.,** Lateral olfactory tract transmitter: glutamate, aspartate or neither?, *Cell. Mol. Neurobiol.*, 1, 115, 1981.
52. **Curtis, D. R.,** Actions of drugs on single neurons in the spinal cord and thalamus, *Br. Med. Bull.*, 21, 5, 1965.
53. **Hosli, L. and Hosli, E.,** Action and uptake of neurotransmitters in CNS tissue culture, *Rev. Physiol. Biochem. Pharmacol.*, 81, 135, 1978.
54. **Logan, W. J. and Snyder, S. H.,** High affinity uptake systems for glycine, glutamic acid and aspartic acid in synaptosomes of rat central nervous tissues, *Brain Res.*, 42, 413, 1972.
55. **Balcar, V. J. and Johnston, G. A. R.,** The structural specificity of the high affinity uptake of L-glutamate and L-aspartate by rat brain slices, *J. Neurochem.*, 19, 2657, 1972.
56. **Balcar, V. J., Borg, J., and Mandel, P.,** High affinity uptake of L-glutamate and L-aspartate by glial cells, *J. Neurochem.*, 28, 87, 1977.
57. **Logan, W. J.,** Amino acid transport by two glial cell lines and by proliferating glia, *Exp. Neurol.*, 53, 431, 1976.
58. **Shank, R. P. and Campbell, G. L.,** Amino acid uptake, content, and metabolism by neuronal and glial enriched cellular fractions from mouse cerebellum, *J. Neurosci.*, 4, 58, 1984.
59. **Gjessing, L. R., Gjesdahl, P., and Sjaastad, O.,** The free amino acids in human cerebrospinal fluid, *J. Neurochem.*, 19, 1807, 1972.
60. **Perry, T. L., Hansen, S., and Kennedy, J.,** CSF amino acids and plasma — CSF amino acid ratios in adults, *J. Neurochem.*, 24, 587, 1975.
61. **Goodnick, P. J., Evans, H. E., Dunner, D. L., and Fieve, R. R.,** Amino acid concentrations in cerebrospinal fluid: effects of aging, depression and probenecid, *Biol. Psychiatry*, 15, 557, 1980.
62. **Ferraro, T. N. and Hare, T. A.,** Free and conjugated amino acids in human CSF: influence of age and sex, *Brain Res.*, 338, 53, 1985.
63. **Shank, R. P., Whiten, J. T., and Baxter, C. F.,** Viability of some metabolic processes in the isolated toad brain adapted to two osmotic environments, *J. Neurobiol.*, 6, 145, 1975.
64. **Shank, R. P. and Baxter, C. F.,** Uptake and metabolism of glutamate by isolated toad brains containing different levels of endogenous amino acids, *J. Neurochem.*, 24, 641, 1975.
65. **Shank, R. P. and Aprison, M. H.,** Glutamine uptake and metabolism by the isolated toad brain: evidence for its proposed role as a transmitter precursor, *J. Neurochem.*, 28, 1189, 1977.
66. **Hamberger, A., Berthold, C. H., Karlsson, B., Lehman, A., and Nystrom, B.,** Extracellular glutamate, GABA and glutamine *in vivo* — perfusion dialysis of the rabbit hippocampus, in *Glutamine, Glutamate and GABA in the Central Nervous System*, Hertz, L., Kvamme, E., McGeer, E., and Schousboe, A., Eds., Alan R. Liss, New York, 1983, 473.
67. **Disbrow, J. K., Gershten, M. J., and Ruth, J. A.,** Uptake of L-[^{3}H] glutamic acid by crude and purified synaptic vesicles from rat brain, *Biochem. Biophys. Res. Commun.*, 108, 1221, 1982.
68. **Naito, S. and Ueda, T.,** Characterization of glutamate uptake into synaptic vesicles, *J. Neurochem.*, 44, 99, 1985.
69. **Krnjevic, K. and Schwartz, S.,** Some properties of unresponsive cells in the cerebral cortex, *Exp. Brain Res.*, 3, 306, 1967.

70. **Hosli, L., Andres, P. F., and Hosli, E.,** Depolarization of cultured astrocytes by glutamate and aspartate, *Neuroscience*, 4, 1593, 1979.
71. **Bowman, C. L. and Kimelberg, H. K.,** Excitatory amino acids directly depolarize rat brain astrocytes in primary culture, *Nature (London)*, 311, 656, 1984.
72. **Tang, C. and Orkand, R. K.,** Glutamate depolarization of glial cells in Necturus optic nerve, *Neurosci. Lett.*, 63, 300, 1986.
73. **Kettenmann, H. and Schachner, M.,** Pharmacological properties of γ-aminobutyric acid-, glutamate-, and aspartate-induced depolarizations in cultured astrocytes, *J. Neurosci.*, 5, 3295, 1985.
74. **Madelian, V., Martin, D. L., Lepore, R., Perrone, M., and Shain, W.,** β-Receptor-stimulated and cyclic adenosine 3′,5′-monophosphate-mediated taurine release from LRM55 glial cells, *J. Neurosci.*, 5, 3154, 1985.
75. **Shank, R. P. and Campbell, G. L.,** Metabolic precursors of glutamate and GABA, in *Glutamine, Glutamate and GABA in the Central Nervous System*, Hertz, L., Kvamme, E., McGeer, E., and Schousboe, A., Eds., Alan R. Liss, New York, 1983, 355.
76. **Berl, S. and Clarke, D. D.,** Metabolic compartmentation of glutamate in the CNS, in *Handbook of Neurochemistry*, Vol. 2, Lajtha, A., Ed., Plenum Press, New York, 1969, 447.
77. **Bradford, H. F., Ward, H. K., and Thomas, A. J.,** Glutamine — a major substrate for nerve endings, *J. Neurochem.*, 30, 1453, 1978.
78. **Szerb, J. C. and O'Regan, P. A.,** Effect of glutamine on glutamate release from hippocampal slices induced by high K^+ or by electrical stimulation: interaction with different Ca^{++} concentrations, *J. Neurochem.*, 44, 1724, 1985.
79. **van den Berg, C. J., Reignierse, G. L. A., Blochuis, G. G. C., Kron, M. C., Ronda, G., Clarke, D. D., and Garfinkel, D.,** A model of glutamate metabolism in brain: a biochemical analysis of a heterogeneous structure, in *Metabolic Compartmentation and Neurotransmission — Relation to Brain Structure and Function*, Berl, S., Clarke, D. D., and Schneider, S., Eds., Plenum Press, New York, 1976, 515.
80. **Shank, R. P. and Aprison, M. H.,** Biochemical aspects of the neurotransmitter function of glutamate, in *Glutamic Acid: Advances in Biochemistry and Phyiology*, Filer, L. J., Jr., Garattini, S., Kare, M. R., Reynolds, W. A., and Wurtman, R. J., Eds., Raven Press, New York, 1979, 139.
81. **Nicklas, W. J.,** Relative contributions of neurons and glia to the metabolism of glutamate and GABA, in *Glutamine, Glutamate and GABA in the Central Nervous System*, Hertz, L., Kvamme, E., McGeer, E., and Schousboe, A., Eds., Alan R. Liss, New York, 1983, 219.
82. **Rothstein, J. D. and Tabakoff, B.,** Alteration of striatal glutamate release after glutamine synthetase inhibition, *J. Neurochem.*, 43, 1438, 1984.
83. **Rothstein, J. D. and Tabakoff, B.,** Glial and neuronal glutamate transport following glutamine synthetase inhibition, *Biochem. Pharmacol.*, 34, 73, 1985.
84. **Thanki, C. M., Sugden, D., Thomas, A. J., and Bradford, H. F.,** *In vivo* release from cerebral cortex of [^{14}C] glutamate synthesized from [U-^{14}C] glutamine, *J. Neurochem.*, 41, 611, 1983.
85. **Shank, R. P. and Aprison, M. H.,** Present status and significance of the glutamine cycle in neural tissues, *Life Sci.*, 28, 837, 1981.
86. **Norenberg, M. D.,** The distribution of glutamine synthetase in the rat central nervous system, *J. Histochem. Cytochem.*, 27, 756, 1979.
87. **Patel, M. S.,** The relative significance of CO_2-fixing enzymes in the metabolism of the rat brain, *J. Neurochem.*, 22, 717, 1974.
88. **Yu, A. C. H., Drejer, J., Hertz, L., and Schousboe, A.,** Pyruvate carboxylase activity in primary cultures of astrocytes and neurons, *J. Neurochem.*, 41, 1484, 1983.
89. **Shank, R. P., Bennett, G. S., Freytag, S. O., and Campbell, G. L.,** Pyruvate carboxylase: an astrocyte-specific enzyme implicated in the replenishment of amino acid neurotransmitter pools, *Brain Res.*, 329, 364, 1985.
90. **Shank, R. P. and Campbell, G. L.,** Avid Na^+-dependent, high affinity uptake of alpha-ketoglutarate by nerve terminal enriched material from mouse cerebellum, *Life Sci.*, 28, 843, 1981.
91. **Shank, R. P. and Campbell, G. L.,** Alpha-ketoglutarate and malate uptake and metabolism by synaptosomes: further evidence for an astrocyte to neuron metabolic shuttle, *J. Neurochem.*, 42, 1153, 1984.
92. **Juggi, J. S., Iyngkaran, N., and Prathap, K.,** Hyperammonemia in Reye's syndrome, in *Reye's Syndrome II*, Crocker, J. F. S., Ed., Grune & Stratton, New York, 1979, 411.
93. **Shank, R. P. and Campbell, G. L.,** Glutamine and alpha-ketoglutarate uptake and metabolism by nerve terminal enriched material from mouse cerebellum, *Neurochem. Res.*, 7, 601, 1982.
94. **Shank, R. P. and Campbell, G. L.,** Glutamine, glutamate and other possible regulators of alpha-ketoglutarate and malate uptake by synaptic terminals, *J. Neurochem.*, 42, 1162, 1984.
95. **Cheng, S. C., Nakamura, R., and Waelsch, H.,** Relative contribution of carbon dioxide fixation and acetyl-CoA pathways in two nervous tissues, *Nature (London)*, 216, 928, 1967.
96. **Shank, R. P. and Campbell, G. L.,** Ornithine as a precursor of glutamate and GABA: uptake and metabolism by neuronal and glial enriched cellular material, *J. Neurosci. Res.*, 9, 47, 1983.

97. **Yoneda, Y., Roberts, E., and Dietz, G. W., Jr.,** A new synaptosomal biosynthetic pathway of glutamate and GABA from ornithine and its negative feedback inhibition by GABA, *J. Neurochem.*, 38, 1686, 1982.
98. **Johnson, J. L. and Roberts, E.,** Proline, glutamate and glutamine metabolism in mouse brain synaptosomes, *Brain Res.*, 323, 247, 1984.
99. **Ratner, S., Morell, H., and Caravalho, E.,** Enzymes of arginine metabolism in brain, *Arch. Biochem. Biophys.*, 91, 280, 1960.
100. **Roberts, E.,** Strategies for identifying sources and sites of formation of GABA-precursor or transmitter glutamate in brain, in *Glutamate as a Neurotransmitter*, DiChiara, G. and Gessa, G. L., Eds., Raven Press, New York, 1981, 91.
101. **Wroblewski, J. T., Blaker, W. D., and Meek, J. L.,** Ornithine as a precursor of neurotransmitter glutamate: effect of canaline on ornithine aminotransferase activity and glutamate content in the septum of rat brain, *Brain Res.*, 329, 161, 1985.
102. **Mangan, J. L. and Whittaker, V. P.,** The distribution of free amino acids in subcellular fractions of guinea pig brain, *Biochem. J.*, 98, 128, 1966.
103. **Rassin, D. K.,** Amino acids as putative transmitters: failure to bind to synaptic vesicles of guinea pig cerebral cortex, *J. Neurochem.*, 19, 139, 1972.
104. **Iansek, R. and Redman, S. J.,** The amplitude, time course and charge of unitary excitatory postsynaptic potentials evoked in spinal motoneurone dendrites, *J. Physiol. (London)*, 234, 665, 1973.
105. **Hertz, L.,** Functional interaction between neurons and astrocytes, *Prog. Neurobiol.*, 13, 177, 1979.
106. **Cupello, A. and Hyden, H.,** A calculation method for evaluating the time course of GABA removal from a synaptic cleft by presynaptic uptake systems, *Brain Res.*, 342, 176, 1985.
107. **Storm-Mathisen, J. and Ottersen, O. P.,** Immunohistochemistry of glutamate and GABA, in *Glutamine, Glutamate and GABA in the Central Nervous System*, Hertz, L., Kvamme, E., McGeer, E., and Schousboe, A., Eds., Alan R. Liss, New York, 1983, 185.
108. **Madl, J. E., Larson, A. A., and Beitz, A. J.,** Monoclonal antibody specific for carbodiimide-fixed glutamate: immunocytochemical localization in the rat CNS, *J. Histochem. Cytochem.*, 34, 317, 1986.
109. **Conti, F., Rustioni, A., and Petrusz, P.,** Morphology and laminar distribution of neurons with glutamate-like immunoreactivity in the rat somatosensory cortex, *Soc. Neurosci. Abstr.*, 11, 755, 1985.
110. **Fagg, G. E. and Foster, A. C.,** Amino acid neurotransmitters and their pathways in the mammalian central nervous system, *Neuroscience*, 9, 701, 1983.
111. **Fonnum, F.,** Glutamate: a neurotransmitter in mammalian brain, *J. Neurochem.*, 42, 1, 1984.
112. **Cuenod, M.,** Identification of pathways for excitatory amino acid transmitters, in *Excitatory Amino Acids*, Roberts, P. J., Storm-Mathisen, J., and Bradford, H. F., Eds., Macmillan, London, 1986, 117.
113. **Oliver, D. L., Potashner, S. J., Jones, D. R., and Morrest, D. K.,** Selective labeling of spiral ganglion and granule cells with D-aspartate in the auditory system of cat and guinea pig, *J. Neurosci.*, 3, 455, 1983.
114. **Hicks, T. P., Ruwe, W. D., Veale, W. L., and Veenhuizen, J.,** Aspartate and glutamate as synaptic transmitters of parallel visual cortical pathways, *Exp. Brain Res.*, 58, 421, 1985.
115. **Kimura, H., Okamoto, K., and Sakai, Y.,** Pharmacological evidence for L-aspartate as the neurotransmitter of cerebeller climbing fibers in the guinea pig, *J. Physiol. (London)*, 365, 103, 1985.
116. **Krnjevic, K.,** Control of neuronal excitability by intracellular divalent cations: a possible target for neurotransmitter actions, in *Neurotransmitter Function — Basic and Clinical Aspects*, Fields, W. S., Ed., Symposia Specialists, Miami, 1977, 11.
117. **Surtees, L. and Collins, G. G. S.,** Receptor types mediating the excitatory action of exogenous L-aspartate and L-glutamate in rat olfactory cortex, *Brain Res.*, 334, 287, 1985.
118. **Zaczek, R., Koller, K. J., and Coyle, J. T.,** N-acetyl-aspartyl-glutamate: regional levels in rat brain and the effects of brain lesions as determined by a new HPLC method, *J. Neurochem.*, 43, 1136, 1984.
119. **French-Mullen, J. M. H., Koller, K., Zaczek, R., Coyle, J. T., Hori, N., and Carpenter, D. O.,** N-acetylaspartylglutamate: possible role as the neurotransmitter of the lateral olfactory tract, *Proc. Natl. Acad. Sci. U.S.A.*, 82, 3897, 1985.
120. **Johnson, J. L.,** Glutamic acid as a synaptic transmitter candidate in the dorsal sensory neuron: reconsiderations, *Life Sci.*, 20, 1637, 1977.
121. **Barbaresi, P., Rustioni, A., and Cuenod, M.,** Retrograde labeling of dorsal root ganglion neurons after injection of tritiated amino acids in the spinal cord of rats and cats, *Somatosensory Res.*, 3, 1, 1985.
122. **Talman, W. T., Perrone, M. H., and Reis, D. J.,** Evidence for L-glutamate as the neurotransmitter of baroreceptor afferent nerve fibers, *Science*, 209, 813, 1980.
123. **Nagg, J. I., Hunt, S. P., Iverson, L. L., and Emson, P. C.,** Biochemical and anatomical observations on the degeneration of peptide-containing primary afferent neurons after neonatal capsaicin, *Neuroscience*, 6, 1923, 1981.
124. **Urban, L., Willetts, J., Randic, M., and Papka, R. E.,** The acute and chronic effects of capsaicin on slow excitatory transmission in the dorsal horn, *Brain Res.*, 330, 390, 1985.
125. **White, S. R.,** A comparison of the effects of serotonin, substance P and thyrotropin-releasing hormone on excitability of rat spinal motoneurons *in vivo*, *Brain Res.*, 335, 63, 1985.

126. **Joels, M. and Urban, I. J. A.,** Arginine[8]-vasopressin enhances the responses of lateral septal neurons in the rat to excitatory amino acids and fimbira-fornix stimuli, *Brain Res.*, 311, 201, 1984.
127. **Gerschenfeld, H. M.,** Chemical transmission in invertebrate central nervous systems and neuromuscular junctions, *Physiol. Rev.*, 53, 1, 1983.
128. **Usherwood, P. N. R.,** Amino acids as neurotransmitters, *Adv. Comp. Physiol. Biochem.*, 7, 227, 1975.
129. **Carew, T. J., Pinsker, H., Rubinson, K., and Kandel, E. R.,** Physiological and biochemical properties of neuromuscular transmission between identified motoneurons and gill muscle in *Aplysia, J. Neurophysiol.*, 37, 1020, 1974.
130. **Carlyle, R. F.,** The occurrence and actions of amino acids on isolated supra oral sphincter preparations of the sea anemone *Actinia equina, J. Physiol. (London)*, 236, 635, 1974.
131. **Shank, R. P. and Freeman, A. R.,** Cooperative interaction of glutamate and aspartate with receptors in the neuromuscular excitatory membrane in walking limbs of the lobster, *J. Neurobiol.*, 6, 289, 1975.
132. **Freeman, A. R., Shank, R. P., Kephart, J., Dekin, M., and Wang, M.,** Modulation of transmission at a glutamate synapse, *J. Physiol. (Paris)*, 75, 605, 1979.
133. **Freeman, A. R., Shank, R. P., Kephart, J., Dekin, M., and Wang, M.,** A model for excitatory transmission at a glutamate synapse, in *Glutamate as a Neurotransmitter*, DiChiara, G. and Gessa, G. L., Eds., Raven Press, New York, 1981, 227.
134. **McCreery, M. J. and Carpenter, D. O.,** Synergistic action of L-glutamate and L-aspartate in Aplysia, *Fed. Proc., Fed. Am. Soc. Exp. Biol.*, 37, 582, 1978.
135. **McBurney, R. N. and Crawford, A. C.,** Amino acid synergism at synapses, *Fed. Proc., Fed. Am. Soc. Exp. Biol.*, 38, 2080, 1979.
136. **Miller, R. J.,** Second messengers, phosphorylation and neurotransmitter release, *Trends Neurosci.*, 8, 463, 1985.

Chapter 16

UPTAKE AND RELEASE OF GLUTAMATE AND GLUTAMINE IN NEURONS AND ASTROCYTES IN PRIMARY CULTURES

Arne Schousboe, Jørgen Drejer, and Leif Hertz

TABLE OF CONTENTS

I. INTRODUCTION

Due to the dual role of glutamate as a neurotransmitter and as an intermediary metabolite, it is of importance to gain knowledge of the dynamic interplay between neurons and astrocytes concerning fluxes of this amino acid in and out of the cells. Since glutamine on one hand is a glutamate metabolite and on the other may serve as an important precursor for glutamate (see Chapter 17), it is equally relevant to discuss uptake and release processes for this amino acid in the different cells types. Such information has been obtained using different kinds of cellular and subcellular (e.g., synaptosome) preparations. A preparation for which detailed information is available is primary cultures of neurons and astrocytes, i.e., cultures obtained directly from the living animals, often at an immature state.[1,2] Such cultures constitute reliable models for their in vivo counterparts and the transport studies can be performed in cells that are undamaged by any separation procedure.[1-4] For this reason, the present review will be restricted to dealing with transport processes in primary cultures of neurons and astrocytes. Cell lines will not be included because of the risk of quantitative alterations in transport phenomena in these transformed cells.[1-3] Initially, a brief description of the cultured cells will be given (Section II).

II. CULTURED CELLS

A. Cerebellar Granule Cells

Biochemical as well as electrophysiological evidence strongly suggest that cerebellar granule cells utilize glutamate as their transmitter.[5-7] Since culture methods for this cell type have been available for some time,[1,2] we have chosen cultured cerebellar granule cells as a model system for glutamatergic neurons.[8-12] From studies of stimulus-coupled release of exogenously supplied L-glutamate or D-aspartate[10,11,13] or endogenously synthesized glutamate,[13,14] it can be concluded that these cells in culture behave as functionally active glutamatergic neurons. A more detailed characterization of these cultures at both the functional and morphological levels can be found elsewhere.[13,15-18] It seems to be essential for functional differentiation of the cells that they have been kept in culture for at least 10 days.[19]

B. Cerebral Cortical Neurons

Recent immunocytochemical studies[20] have indicated that a large fraction of cerebral cortical interneurons (stellate cells) are GABAergic.[1,2] Methods for cultivation of cortical interneurons have also been available for several years. Numerous studies of transmitter biosynthesis, uptake, and release as well as of electrophysiology in such cultures strongly suggest that the majority of the neurons in these cultures are indeed GABAergic.[21-27] This culture system, which is described in detail elsewhere,[1,2,9,25,27,28] has therefore been employed as a model system for GABAergic neurons. These cells also need to be kept in culture for at least 10 days to achieve optimum maturation.[19]

C. Astrocytes

It is by now widely accepted that astrocytes play an active role in the maintenance of a normal extracellular milieu of ions as well as of neurotransmitters in the CNS.[29] Protocols for primary cultures of astrocytes have been available for more than a decade,[1,2,4] and biochemical and functional characterization of these cultures has suggested that the cultured astrocytes may be well suited as a model system for their in vivo counterparts.[29] Although some evidence points to the existence of a functional heterogeneity of astrocytes originating from different brain regions,[30-35] we have mainly used astrocytes cultured from cerebral cortex[4] for the studies of glutamate and glutamine transport. Also, these cells mature in culture and should be maintained for a minimum of 3 weeks before their use.[35]

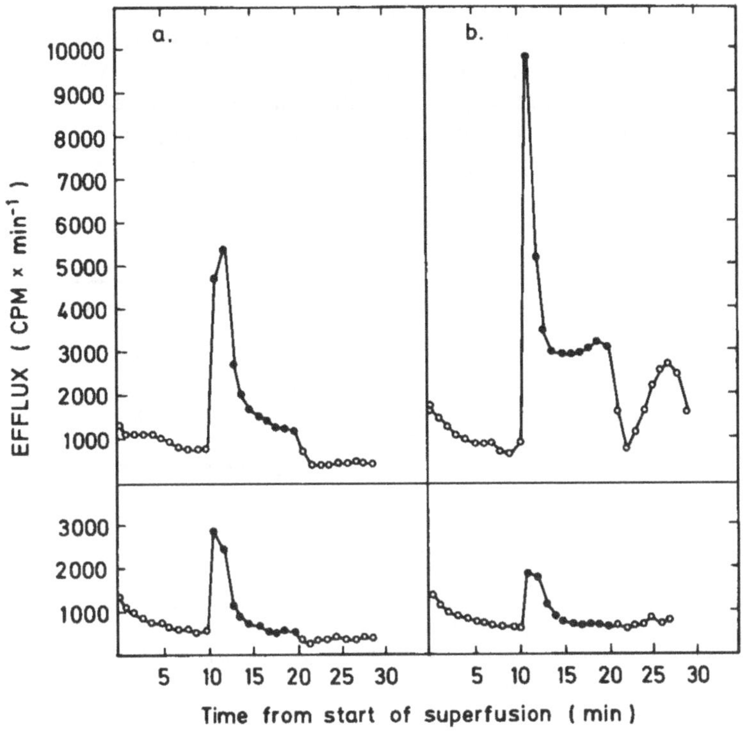

FIGURE 1. Efflux of radioactivity (cmp × min^{-1}) from cultured cerebellar granule cells preloaded with D-[^{3}H]aspartate as a function of time from the start of superfusion. After 10 min of superfusion, the medium was changed from a physiological medium to a corresponding medium containing 55 m*M* KCl (a, upper part) or 55 m*M* KCl without glucose and equilibrated with N_2 (b, upper part). After an additional 10 min, the high-potassium media were switched back to the physiological medium. The periods of elevated K^+ are indicated by closed symbols. In the lower parts of the figures, the superfusion media were analogous to the above-mentioned media, but with $CaCl_2$ substituted by $CoCl_2$ (1.0 m*M*) during 8 to 20 min from the start of superfusion. (From Drejer, J., Benveniste, H., Diemer, N. H., and Schousboe, A., *J. Neurochem.*, 45, 145, 1985. With permission.)

III. GLUTAMATE

A. Release

1. Neurons

a. Cerebellar Granule Cells

Endogenous as well as exogenously supplied glutamate or its nonmetabolizable analogue D-aspartate is released from cultured cerebellar granule cells in response to elevated potassium concentrations in a calcium-dependent manner,[10,11,13,14,17,36-38] as illustrated in Figure 1a. The stimulated release amounts to about 10 nmol × min^{-1} × mg^{-1} (Table 1 and Figure 2), which is many times higher than the release rate observed under resting conditions, i.e., at 5 m*M* KCl (Table 1). An enhancement of the release can also be obtained by either L-glutamate, L-aspartate, or D-aspartate (Figure 3). Other glutamate agonists such as quisqualate, *N*-methyl-D-aspartate (NMDA), and kainate do not exert such an effect.[41] This might suggest that the stimulation could be a result of a homoexchange process. However, the facts that the stimulated release is strictly Ca^{2+}-dependent and that the glutamate antagonists γ-D-glutamylglycine, glutamate diethylester, D-aminophosphonovaleric acid, and D-α-aminoadipate block the L-glutamate-induced transmitter release strongly argue against a hom-

Table 1
RATES OF RELEASE OF GLUTAMATE (GLU) AND GLUTAMINE (GLN) FROM PRIMARY CULTURES OF CEREBELLAR GRANULE CELLS, CEREBRAL CORTICAL NEURONS, AND CEREBRAL CORTICAL ASTROCYTES

		Release rate ($nmol \times min^{-1} \times mg^{-1}$)	
Cell type	**Amino acid**	**Basal**	**Stimulated**[a]
Granule cells	Glu		4.7[10]
		1.7[10]	13.0[36]
Cerebral neurons	Glu	1.3[27]	0.8[27]
Astrocytes	Glu	1.8[39]	
		1.3—2.6[10]	4—8[36]
Granule cells	Gln	0.2[b,75]	
Cerebral neurons	Gln	1.0[27]	
Astrocytes	Gln	2.0[40]	

[a] Measured as the total release rate into a medium with elevated potassium concentration.
[b] Measured in the presence of the glutaminase inhibitor 6-diazo-5-oxo-L-norleucine.

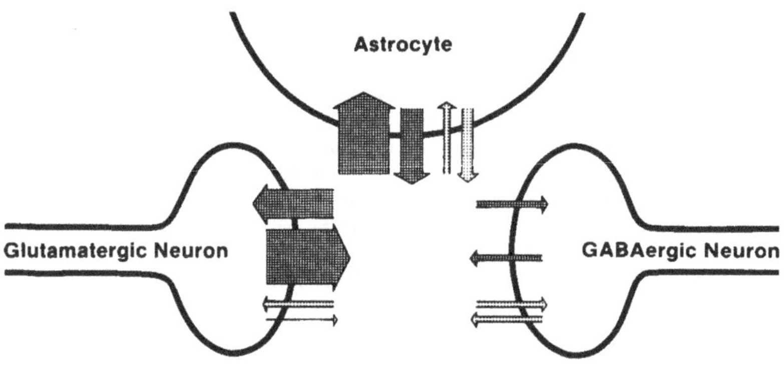

FIGURE 2. Schematic drawing of release and uptake of glutamate (▬) and glutamine (:::) from glutamatergic or GABAergic neurons and astrocytes. The release of glutamate from the glutamatergic neurons corresponds to that observed during excitation. The sizes of the arrows give an estimate of the relative magnitudes of the respective fluxes. It can be seen that the majority of the neuronally released glutamate is accumulated into astrocytes. Moreover, it should be noted that very little glutamine is being released from glutamatergic neurons.

oexchange mechanism. The conclusion therefore is that cerebellar granule cells possess a glutamate receptor, which is coupled to transmitter release and has unique pharmacological properties.[41] Recent autoradiographic studies of glutamate binding in cerebellar slices have indicated the presence of a subpopulation of glutamate receptors in the granule layer which show similar pharmacological characteristics.[42]

The potassium-stimulated glutamate release from cerebellar granule cells has recently been reported to be regulated by the inhibitory amino acid neurotransmitter GABA.[13,43,44] Under conditions where the neurons have both high- and low-affinity GABA receptors,[13,43-45] GABA

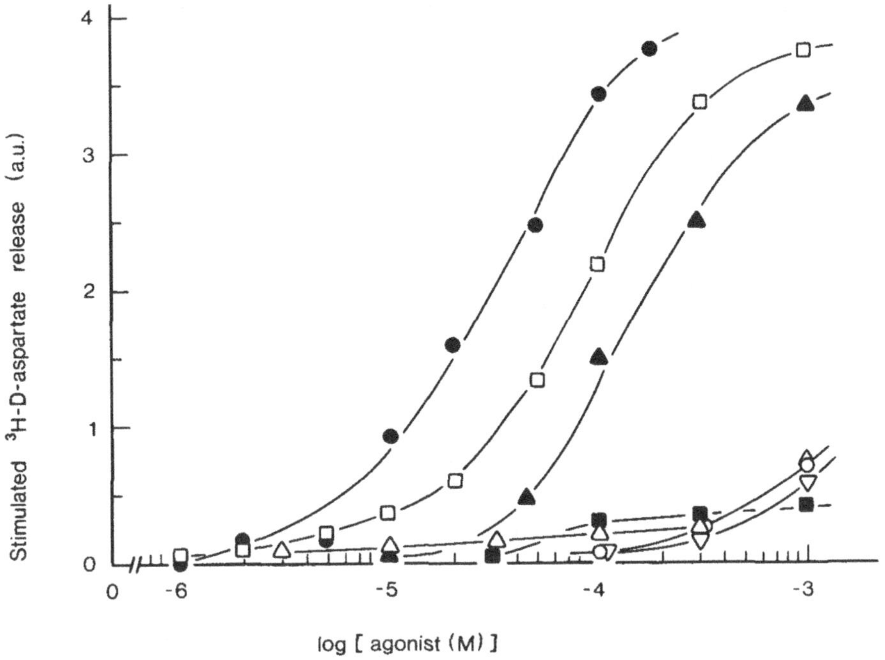

FIGURE 3. Stimulation of release of ^{3}H-D-aspartate from cultured cerebellar granule cells preloaded with ^{3}H-D-aspartate. Cells were exposed to increasing concentrations of L-glutamate (●), L-aspartate (□), D-aspartate (▲), quisqualate (△), NMDA (■), kainate (○), and L-α-aminoadipate (▽) as the stimulating agent. The stimulated release represents the release in the presence of the agonist corrected for the released observed in Ca^{2+}-free media. (From Drejer, J., Honore, T., Meier, E., and Schousboe, H., *Life Sci.*, 38, 2077, 1986. With permission.)

has been found to inhibit evoked glutamate release.[13] This effect of GABA could be mimicked by its agonists muscimol and THIP (4,5,6,7-tetrahydroisoxazolo[5,4-*c*]pyridin-3-ol) and blocked by its antagonist bicuculline.[13] This together with the finding that the $GABA_B$-receptor agonist baclofen[46] had no effect[13] suggest that the inhibitory effect of GABA on these neurons is mediated by a $GABA_A$ receptor.

b. Cerebral Cortical Neurons

The release of exogenously supplied glutamate from cerebral cortical, mainly GABAergic, neurons (Table 1 and Figure 2), is smaller than that observed in the cerebellar granule cells and it is not enhanced by excess potassium.[24,27] Moreover, no potassium-stimulated release of exogenously supplied glutamate could be demonstrated from mixed astrocytic-neuronal cultures from chick embryo cerebral cortex.[47]

2. *Astrocytes*

The release of glutamate from astrocytes amounts to about 2 nmol $\times$ min^{-1} $\times$ mg^{-1} (Table 1 and Figure 2). A pronounced potassium-stimulated glutamate release also occurs from astrocytes (Figure 4a), but in contrast to the release of transmitter glutamate from glutamatergic neurons, the K^+-stimulated release from astrocytes is not calcium dependent.[36] This finding is in contrast to previous reports of an absence of a potassium-induced stimulation of glutamate release from astrocytes.[10,14,47] This discrepancy may be explained by the fact that the older experiments were performed under conditions where the potassium-stimulated uptake of glutamate (see Section III.B.2) might have counteracted the increase in release rate.

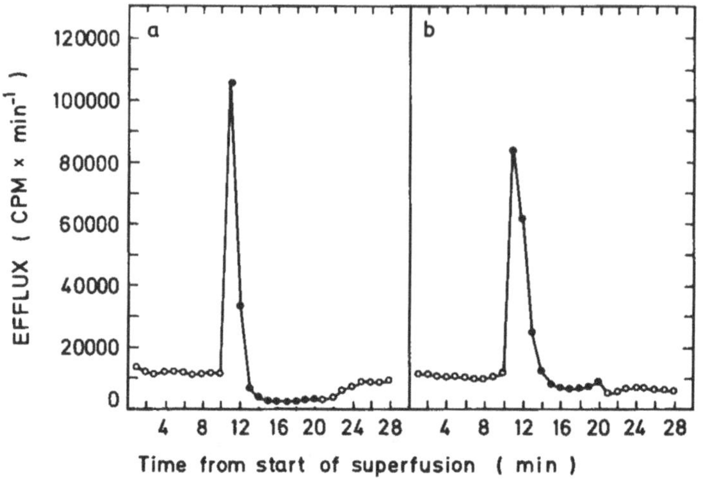

FIGURE 4. Efflux of radioactivity (cpm $\times$ min^{-1}) from cultured astrocytes preloaded with D-[^{3}H]aspartate as a function of time from the start of superfusion. A physiological medium was used as the superfusion medium except for the periods indicated by closed symbols. During these periods, a modified physiological medium was used as superfusion medium containing 55 m*M* KCl (a) or 55 m*M* KCl without glucose and N_2 equilibrated (b). (From Drejer, J., Benveniste, H., Diemer, N. H., and Schousboe, A., *J. Neurochem.*, 45, 145, 1985. With permission.)

B. Uptake

1. Neurons

a. Cerebellar Granule Cells

K_m and V_{max} values for the uptake of glutamate into granule cells are shown in Table 2. From these values the uptake rates were calculated at 10 μ*M* which seems to be the physiological extracellular concentration of glutamate.[52,53] It can be seen (Table 2) that the uptake rate is 2 to 8 nmol $\times$ min^{-1} $\times$ mg^{-1}. This is lower than the release rate (Table 1 and Figure 2). This difference becomes more evident when uptake and release rates from the same study are compared, i.e., a release rate of 4.7 (Table 1) and uptake rate of 2.2 nmol $\times$ min^{-1} $\times$ mg^{-1} (Table 2) observed by Drejer et al.[10]

The glutamate uptake into cerebellar granule cells is strictly dependent upon the presence of sodium.[10] To determine the number of sodium ions transported per molecule of glutamate, plots of 1/v vs. 1/Na^+ or (1/Na^+)^2 were constructed and straight lines could be obtained only in the latter plot (Figure 5), indicating that two sodium ions are cotransported with each molecule of glutamate. Since the glutamate uptake probably is energetically driven by the sodium gradient, this enables the cells to establish a high ratio between intracellular and extracellular glutamate.[54] The uptake of glutamate is also sensitive to potassium since it is doubled in the presence of 50 m*M* KCl.[36] Nevertheless, the potassium-induced release (Table 1) is considerably larger than the potassium-stimulated uptake.

The glutamate carrier is not specific for L-glutamate, but is shared by L- and D-aspartate which are transported with essentially identical K_m and V_{max} values.[11] This is further underlined by results of a study of the mutual inhibition pattern of the uptake of the three amino acids (Table 3). It can be seen that the K_m and K_i values are identical which clearly shows that L-glutamate, L-aspartate, and D-aspartate share the same carrier.[11] Based on studies of the inhibition of glutamate uptake with different glutamate analogues, such as 3-hydroxyaspartate, D-glutamate, the β- and γ-hydroxamic acid derivatives of D- and L-aspartate and L-glutamate, and cysteinate and homocysteinate in different preparations,[10,11,55-59] a model

Table 2
RATES (NMOL × MIN^{-1} × MG^{-1}) OF UPTAKE OF GLUTAMATE (GLU) AND GLUTAMINE (GLN) INTO PRIMARY CULTURES OF CEREBRAL GRANULE CELLS, CEREBRAL CORTICAL NEURONS, AND CEREBRAL CORTICAL ASTROCYTES

	Glu			Gln		
Cells	K_m (μM)	V_{max}	$V_{10\mu M}$[a]	K_m (mM)	V_{max}	$V_{100\mu M}$[a]
Granule cells	42	10.2	2.2[10]	0.7	10.3	1.3[8]
	24	23.0	8.0[9]			1.0[9]
Cortical neurons	43	5.9	1.2[10]	3.0	28.2	0.9[8]
	36	9.0	2.6[9]	3.0	1.0	0.5[48]
	34	0.6	0.2[48]			0.7[9b]
			0.2[49c]			
Astrocyte cortex	63	47.0	7.0[50]	0.2	2.0	0.8[51]
	50	59.0	10.8[39]	0.2	2.0	0.7[48]
	10	4.0	2.6[51]	3.3	50.2	1.5[40]
	4	2.2	2.4[48]			

[a] Unless otherwise indicated, calculated from the V_{max} and K_m values by aid of the Michaelis-Menten equation.

[b] Measured at 100 μM.

[c] Calculated from the uptake at 35 μM on the assumption of a K_m of about 35 μM.

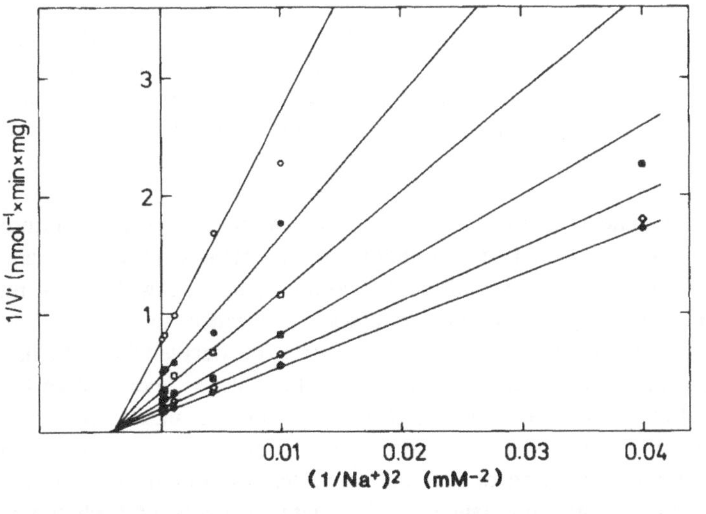

FIGURE 5. Double reciprocal plots of the velocities of L-glutamate uptake into cerebellar granule cells vs. $1/(Na^+)^2$ at six different glutamate concentrations: (○) = 15, (●) = 30, (□) = 50, (■) = 100, (◇) = 100, and (◆) = 1000 μM. Each experimental point represents the average of five or six individual experiments. (From Drejer, J., Larsson, O. M., and Schousboe, A., *Exp. Brain Res.*, 47, 259, 1982. With permission.)

Table 3
CHARACTERIZATION OF UPTAKE OF D- AND L-ASPRATATE AND L-GLUTAMATE INTO CEREBELLAR GRANULE CELLS AND ASTROCYTES

	Inhibiting amino acid					
	Neurons			Astrocytes		
Transported amino acid	D-Asp	L-Asp	L-Glu	D-Asp	L-Asp	L-Glu
D-Asp	26[a]	18	21	83[a]	82	75
L-Asp	29	32[a]	31	70	77[a]	73
L-Glu	40	40	20[a]	106*	68	67[a]

Note: The table gives K_m values (μM, identified by "a") for uptake of the amino acids and K_i values (μM) for their mutual inhibition; * indicates mixed competitive/noncompetitive inhibition; where not otherwise mentioned, the kinetic pattern conforms to linear competitive inhibition.[11]

From Drejer, J., Larsson, O. M., and Schousboe, A., *Neurochem. Res.*, 8, 231, 1983. With permission.

of the glutamate carrier has been proposed.[60] As can be seen in Figure 6, this model has three binding sites for glutamate, one for each of the three charged groups in L-glutamate. The binding site can accommodate the substrates L-glutamate, D- and L-aspartate, L-aspartate-β-hydroxamate, cysteinate, and threo-3-hydroxy-D- or L-aspartate, but not, for example, D-glutamate, L-α-aminoadipate, and the erythro-3-hydroxyaspartates L-glutamate-γ-hydroxamate, D-aspartate-β-hydroxamate, and homocysteinate mainly due to steric hindrance at the binding site for the ω-carboxyl group.

b. Cerebral Cortical Neurons.

From the results presented in Table 2, it can be seen that both the V_{max} for glutamate uptake and the velocity for glutamate uptake at 10 μM glutamate are distinctly lower in cerebral cortical neurons than in cerebellar granule cells (Figure 2). The uptake of about 1 to 2 nmol $\times$ min^{-1} $\times$ mg^{-1} at 10 μM almost equals the release rate (Table 1).

2. Astrocytes

An intense glutamate uptake has consistently been found into primary cultures of astrocytes (Table 2 and Figure 2).* It can be seen that the uptake rates at 10 μM glutamate are at least as high as those in cerebellar granule cells and the values are well above the release rates for glutamate (Table 1). Quite pronounced regional differences in astrocytic glutamate uptake are found.[10,31,34] Thus, the uptake is highest in astrocytes derived from the neostriatum, i.e., a region with a high glutamatergic input. In addition, it has been shown that treatment of the astrocytes during the culturing period with conditioned media from different types of neurons causes an increase in the uptake rate of glutamate.[33] The ability of neuron-exerted factors to induce glutamate uptake sites on astrocytes was found to be most pronounced for neurons which express an abundance of glutamate receptors of both the quisqualate, *N*-methyl-D-aspartate and kainate types.[33,61,62] This would suggest that the capacity for glutamate uptake in astrocytes is regulated by neurons which receive a strong glutamatergic input. If this is indeed the case, it underlines the functional importance of astrocytes for the inactivation of neurotransmitter glutamate by high-affinity uptake. Studies aimed at the

* A lower value for glutamate uptake into astrocytes in primary cultures obtained by Schousboe et al.[59] has been excluded because the uptake was measured into cells that had been detached from the dishes before the measurements.

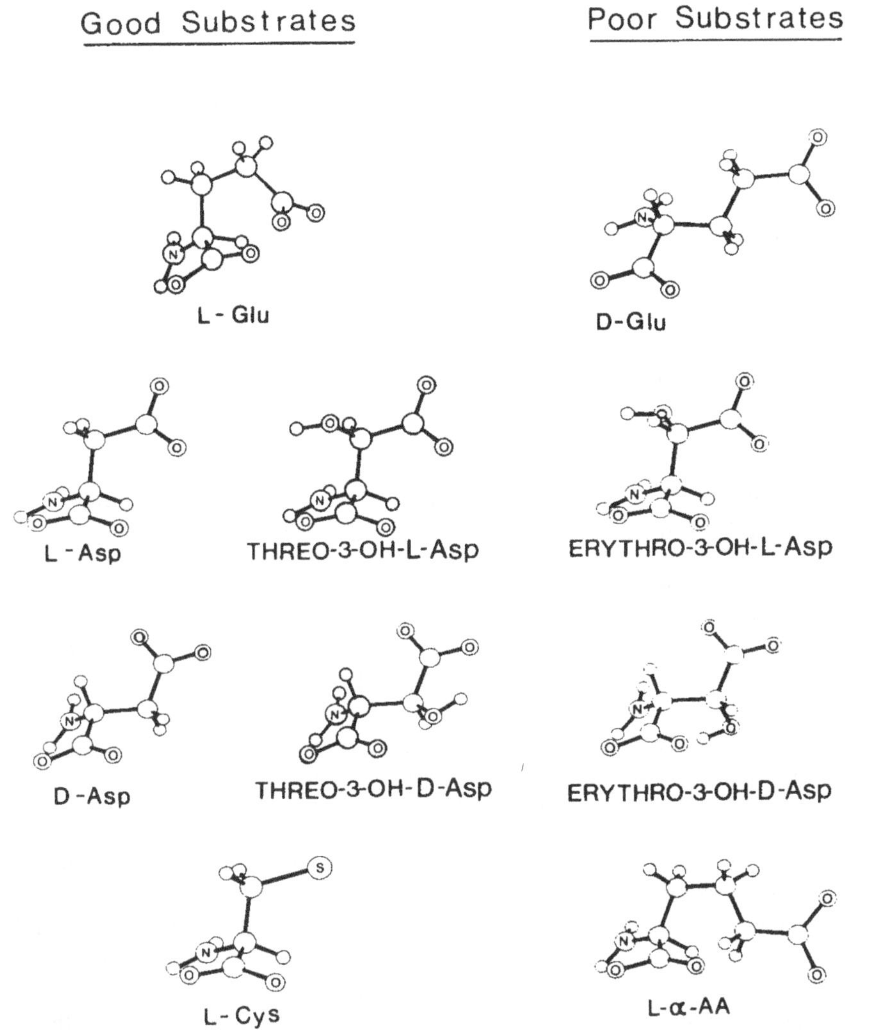

FIGURE 6. On the bases of studies of uptake or inhibition of uptake of L-glutamate and corresponding structures,[10,11,55-60] it may be possible to make a tentative model of the L-glutamate molecule in which it recognizes the uptake site. L-Glutamate and corresponding structures have been drawn in a configuration in which carbon atoms 2, 3, and 4 are in one plane (the plane of the paper). The carboxylic group (C_1) is situated above this plane which leaves the amino group below the plane. It seems that these two amino acid functions, the acid group (e.g., COOH) and the basic group (e.g., NH_2), are essential for the binding to the uptake carrier (e.g., the dicarboxylic acid glutarate shows no affinity for the uptake site). It also seems to be essential that there is an acid function in the other terminal of the molecule; this could be, for example, -COOH, -SH, -OSO_2OH, or -$COHN_2OH$. Substances without this acid function (e.g., GABA and glutamine) do not bind to the uptake site. The finding that L-glutamate and not D-glutamate shows high affinity for the uptake site together with the observation that both L- and D-aspartate are good substrates indicate that the uptake site recognizes L-glutamate in a folded conformation as indicated in the figure. The acid functions of L- and D-aspartate are easily brought into the same area as in L-glutamate in its folded conformation, whereas this is not the case for D-glutamate. When other more bulky acid functions such as -SH, -O-SO_2OH, and -$CONH_2OH$ are substitited for the -COOH group, only the L-configuration appears to show affinity for the uptake site, e.g., L-cysteate but not D-cysteate and L-aspartate-β-hydroxamate but not the D-configuration will inhibit L-glutamate uptake). Also, in substances with an elongated carbon chain, e.g., L-α-aminoadipate (L-αAA), the affinity for the uptake site is lost. The most potent inhibitors of L-glutamate uptake are the two threo isomers of 3-hydroxy-aspartate.[58] It is interesting to note that the erythro forms show very low affinity indicating a sterical hindrance below the plane of the paper. (The drawings for this figure were made by the aid of the ChemGraph program. The assistance by Dr. I. Peterson, Royal Danish School of Pharmacy, in the computer drawings is gratefully acknowledged.)

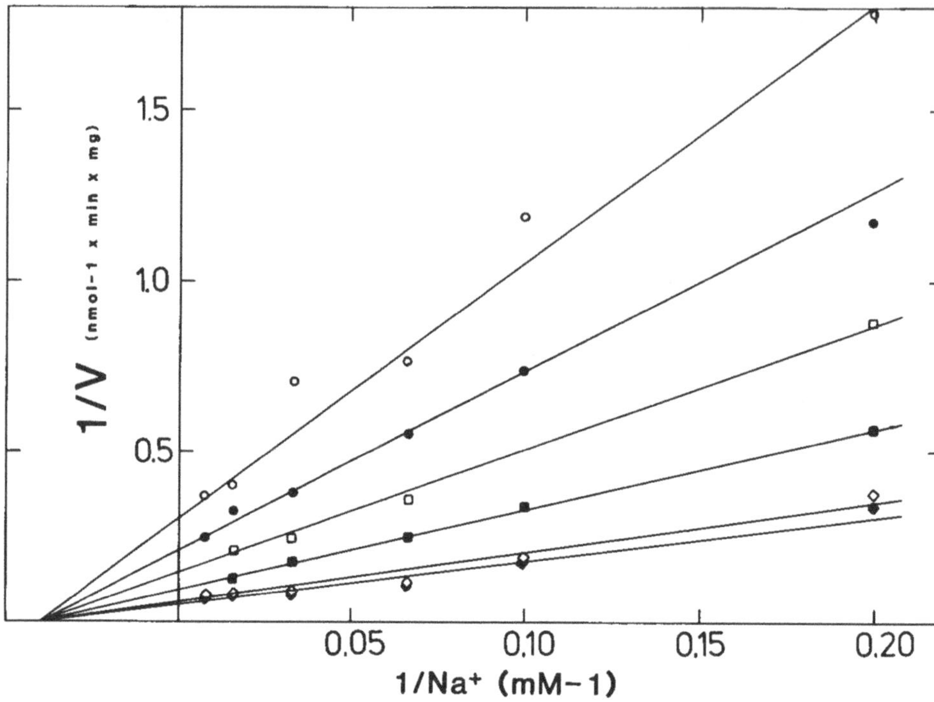

FIGURE 7. Double reciprocal plots of the velocities of L-glutamate uptake into astrocytes vs. 1/Na^+ at six different glutamate concentrations: (○) = 15, (●) = 30, (□) = 50, (■) = 100, (◇) = 300, and (◆) = 1000 μ*M*. Each experimental point represents the average of five or six individual experiments.[75]

characterization of the neuronal factors responsible for this induction of astrocytic glutamate uptake sites are now in progress.[61]

In contrast to cerebellar granule cells, the uptake of glutamate into astrocytes is coupled to only one sodium ion as illustrated in Figure 7. This means that the maximum possible ratio between intracellular and extracellular glutamate contents will be lower than in cerebellar granule cells where glutamate is cotransported with two molecules of sodium. Nevertheless, astrocytes are able to maintain high intracellular glutamate concentrations.[63,64] The glutamate uptake is also dependent upon the potassium concentration since it is stimulated by increased potassium levels in the range of 10 to 50 m*M*.[36,59] Recently, it has also been reported that part of the glutamate uptake into astrocytes is chloride dependent.[65]

From studies of the substrate specificity of the astrocytic glutamate carrier,[10,11,37,59] it is apparent that small differences exist between the neuronal and the astrocytic glutamate carrier, particularly at the binding site for the ω-carboxyl group (Figure 6). This explains why D-aspartate does not behave as a simple competitive inhibitor of L-glutamate uptake in astrocytes (Table 3). This difference between astrocytes and neurons is further illustrated in Figure 8a and b, which shows double reciprocal plots of glutamate uptake into neurons and astrocytes as a function of the glutamate concentration in the absence and presence of D-aspartate. It is seen that while the inhibition in the case of neurons is competitive (i.e., intersection on the ordinate), it is of a mixed type in the case of astrocytes (i.e., intersection to the left of the ordinate).

C. Effect of Ischemia on Glutamate Release and Uptake

1. Cerebellar Granule Cells

Pathological conditions may influence the potassium-induced, calcium-dependent glutamate release. It thus becomes 2.5 times larger under combined anoxia and glucose deprivation

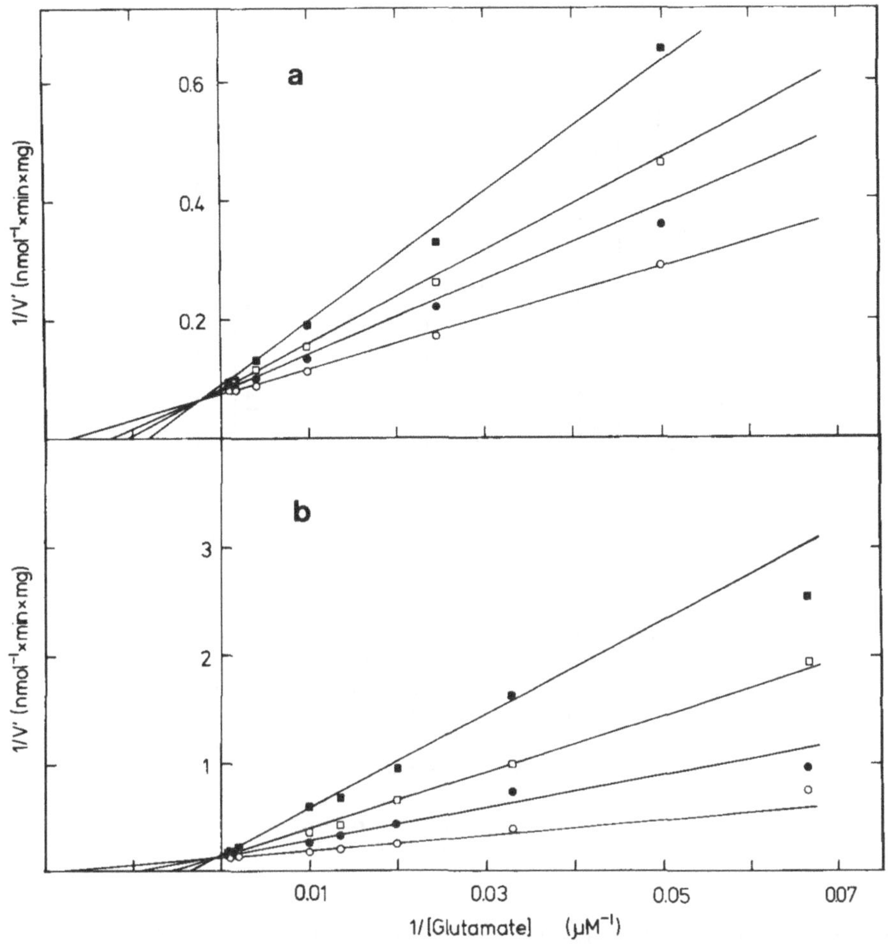

FIGURE 8. Double reciprocal plots of initial L-glutamate uptake rates into astrocytes cultured from prefrontal cortex (panel a) and into cerebellar granule cells (panel b) vs. L-glutamate concentrations. The points represent reciprocal velocities of L-glutamate uptake in the absence (○) and presence of (Panel a) 40 (●), 80 (□), and 160 μM (■) or (panel b) 50 (●), 100 (□), and 200 μM (■) D-aspartate. Each experimental point represents the average of five to eight individual determinations, and the lines were fitted to the experimental points by weighted regression analysis. The inhibitory pattern in panel a conforms to intersecting, linear, noncompetitive inhibition. The reciprocal form of the velocity equation for this type of inhibition is

$$\frac{1}{v} = \frac{K_m}{V_{max}}\left(1 + \frac{I}{K_i}\right)\frac{1}{S} + \frac{1}{V_{max}}\left(1 + \frac{I}{\alpha \cdot K_i}\right)$$

From the lines presented in this panel, the parameters K_i and α can be calculated to be 106 and 8.1 μM, respectively. (From Drejer, J., Larsson, O. M., and Schousboe, A., *Exp. Brain Res.*, 47, 259, 1982. With permission.)

(Figure 1a and b), i.e., a condition mimicking that in the brain in vivo during ischemia. This phenomenon is likely to contribute substantially to the increase in extracellular glutamate concentration which is observed in vivo under ischemia.[53] It is in agreement with this concept that application of the calcium antagonist cobalt completely prevents the effect of ischemia on glutamate release in vivo.[36] Moreover, the potassium-stimulated uptake of glutamate is severely reduced by anoxia and glucose deprivation either separately or in combination

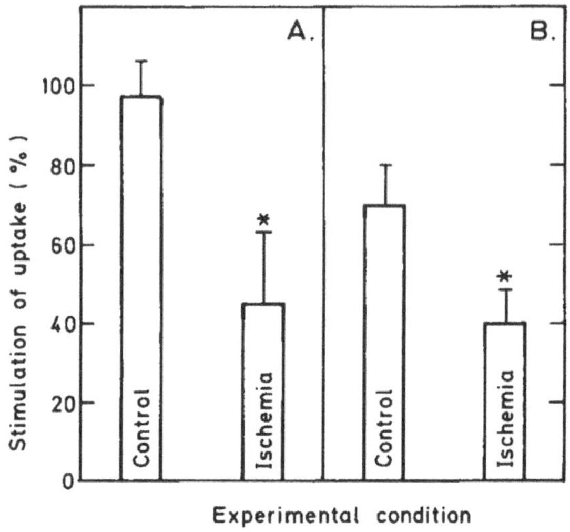

FIGURE 9. Uptake rates (nmol × min^{-1} × mg^{-1} cell protein) for D-[^{3}H]aspartate in cultured cerebellar granule cells (A) and in astrocytes (B) at 55 m*M* KCl as percent stimulation of the uptake rate at normal (5 m*M*) KCl. The incubation medium was a physiological HEPES-buffered saline containing 55 m*M* KCl in the control experiment and a corresponding medium without glucose and equilibrated with N_2 in the experiment called "ischemia". The columns represent the average ± SEM of six individual experiments. Asterisks indicate statistical significant differences from the control experiments ($p<0.05$). (From Drejer, J., Benveniste, H., Diemer, N. H., and Schousboe, A., *J. Neurochem.*, 45, 145, 1985. With permission.)

(Figure 9A). The sensitivity to glucose deprivation is in agreement with the report that granule cells are virtually unable to utilize glutamate as an alternative metabolic substrate[66] (see also Table 2 in Chapter 17).

2. Astrocytes

In contrast to the potassium-induced glutamate release from cerebellar granule cells, the potassium-induced (calcium-independent) release of glutamate from astrocytes is unaffected by combined anoxia and lack of glucose (Figure 4a and b). The potassium-stimulated uptake of glutamate is, however, reduced by anoxia with or without simultaneous glucose deprivation (Figure 9B). In contrast to what was found for cerebellar granule cells, it is not affected by glucose deprivation alone which is in agreement with the finding that astrocytes are capable of utilizing other substrates than glucose, e.g., glutamate and glutamine, as energy sources.[63,67]

IV. GLUTAMINE

A. Release

1. Neurons

a. Cerebellar Granule Cells

The rate of glutamine release from cerebellar granule cells preloaded with 0.5 m*M* glutamine has been measured in the presence of the glutaminase inhibitor 6-diazo-5-oxo-L-norleucine[68] and found to be 0.2 nmol × min^{-1} × mg^{-1},[75] i.e., a remarkably low value (Table 1 and Figure 2).

b. Cerebral Cortical Neurons

Yu et al.[27] have reported a release rate of 1 nmol × min^{-1} × mg^{-1} for cerebral cortical neurons (Table 1 and Figure 2). This value may, however, not be strictly comparable with that in the granule cells since no glutaminase inhibitor was present in these experiments. From Ramaharobandro et al.,[48] a release of approximately similar magnitude can be estimated but no exact values can be calculated.

2. Astrocytes

A release of glutamine from astrocytes has been described by Schousboe et al.[40] and Ramaharobandro et al.[48] The release rate is somewhat higher than in neurons and amounts (in the absence of a glutaminase inhibitor) to about 2 nmol × min^{-1} × mg^{-1} (Table 1 and Figure 2).

B. Uptake

1. Neurons

a. Cerebellar Granule Cells

The K_m and V_{max} values are shown in Table 2 together with the uptake rate at 100 μ*M* glutamine.[8,9] This concentration is close to the extracellular glutamine concentration in the CNS in vivo measured by the microdialysis technique.[52,53] The uptake rates at 100 μ*M* are similar to those in cortical neurons and in astrocytes (Table 2). They are well above the release rate from the granule cells (Figure 2).

b. Cerebral Cortical Neurons

From Table 2 it can be seen that the uptake rates at 100 μ*M* glutamine into cerebral cortical neurons are similar in the studies by Hertz et al.[8] and Ramaharobandro et al.,[48] although considerable differences are found in K_m and V_{max} values. Again the uptake rate at 100 μ*M* is comparable to the release rate (Figure 2).

2. Astrocytes

Also in the case of astrocytes, the K_m and V_{max} values differ between different studies (Table 2), but the uptake velocity at 100 μ*M* glutamine has consistently been found to be 1 to 2 nmol × min^{-1} × mg^{-1}.[39,48,51] This is similar to the release rate (Figure 2).

V. γ-GLUTAMYLTRANSFERASE AND CELLULAR UPTAKE OF GLUTAMATE AND GLUTAMINE

γ-Glutamyltransferase (γ-glutamyltranspeptidase, γ-GT) is a membrane-bound glycoprotein which has been found in brain tissue,[69] synaptosomes,[70] and primary cultures of neurons and astrocytes.[71,72] Since γ-GT has been associated with cellular uptake of glutamine and glutamate,[69,70] it seemed of interest to compare γ-GT activity with the content of glutamate during the development of primary cultures of cerebellar granule cells and cerebral cortical neurons. Figure 10 shows that in both cell types there is a remarkably good correlation between contents of glutamate and γ-GT activity.[72] It can be seen that the activity (Figure 10 and Chapter 17) amounts to 1 to 2 nmol × min^{-1} × mg^{-1}, i.e., a value which is comparable to the uptake of glutamine, a precursor for glutamate in the cells (Chapter 17), but considerably lower than the uptake rates for glutamate. Moreover, these studies were carried out in the absence of glutamate, but in the presence of glutamine in the medium, suggesting that the glutamate content in the cells to a large extent might be maintained by the accumulation of glutamine.

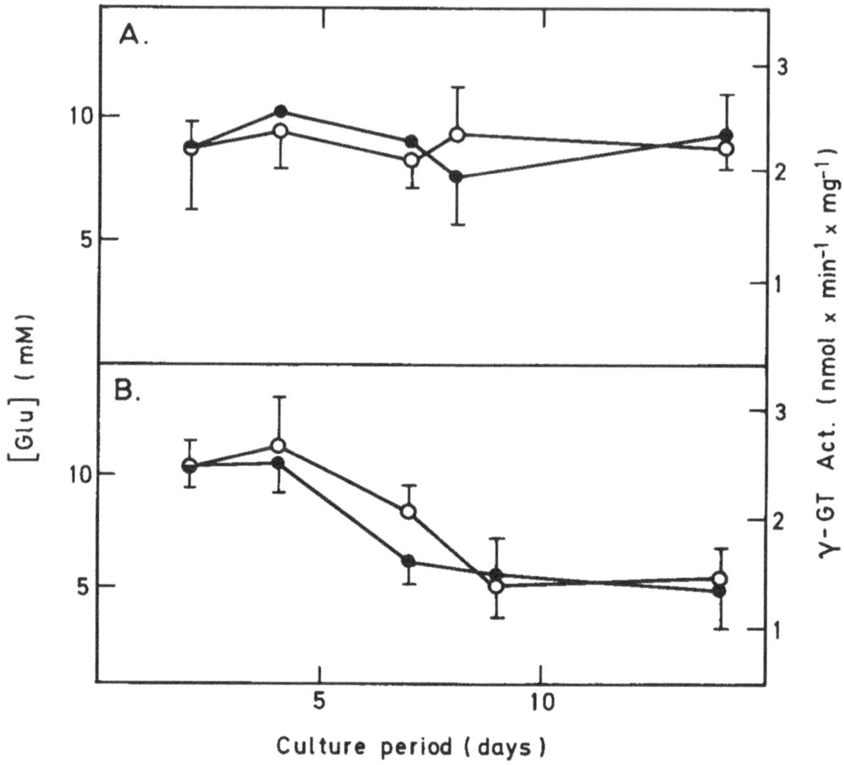

FIGURE 10. Specific activity of γ-GT (●), and cellular content of glutamate (○) in cultured cerebellar granule cells (A) and cultured cerebral cortex neurons (B) as a function of the culture period. Bars indicate SEM values. (From Kvamme, E., Schousboe, A., Hertz, L., Torgner, I. Aa., and Svenneby, G., *Neurochem. Res.*, 10, 993, 1985. With permission.)

VI. CONCLUDING REMARKS

When rates of uptake and of release of glutamate and glutamine in cerebellar granule cells, cerebral cortical neurons, and astrocytes are considered together (Tables 1 and 2), it becomes obvious that during neuronal function there must be a considerable transfer of glutamate from neurons, and especially glutamatergic neurons, such as cerebellar granule cells, to astrocytes (Figure 2). This is in agreement with a pronounced glial accumulation of glutamate in the brain in vivo.[73] Some, but far from all, of this glutamate is converted to glutamine in the astrocytes (Chapter 17). Moreover, the rate of release of glutamine from astrocytes is higher than that from neurons and especially from the glutamatergic cerebellar granule cells which release very little glutamine (Figure 2). Nevertheless, the uptake rate for glutamine in neurons is less than the uptake rate of glutamate in astrocytes (Table 2 and Figure 2) and glutamine is accumulated at similar rates in astrocytes and neurons. Thus, although some of the glutamate which is transferred from neurons to astrocytes may be converted to glutamine, released from the astrocytes, and accumulated into the neurons where it may serve as a precursor for glutamate (and GABA), this mechanism is not able to quantitatively replenish the glutamate pools in neurons.[8,74] Additional glutamate precursors must therefore exist (Chapter 15). This conclusion is supported by studies of actual metabolic fluxes and its possible implications for brain function are discussed in Chapter 17.

Another conclusion to be drawn from the present work is that adverse conditions such as anoxia and/or glucose deprivation severely affect glutamate release and uptake. This may to a large extent contribute to the detrimental effect of ischemia on central nervous tissue. Thus, homeostatic mechanisms for maintenance of intra- and extracellular glutamate concentrations seem to play a major role for the function of the CNS under both normal and abnormal conditons.

REFERENCES

1. **Hertz, L., Juurlink, B. H. J., and Szuchet, S.,** Cell cultures, in *Handbook of Neurochemistry,* Vol. 8, 2nd ed., Lajtha, A., Ed., Plenum Press, New York, 1985, 603.
2. **Hertz, L., Juurlink, B. H. J., Szuchet, S., and Walz, W.,** Cell and tissue culture, in *Neuromethods,* Vol. 2, Boulton, A. A. and Baker, G. B., Eds., Humana Press, Clifton, N.J., 1985, 117.
3. **Schousboe, A.,** Differences between astrocytes in primary culture and glial cell lines in uptake and metabolism of putative amino acid transmitters, in *Cell, Tissue and Organ Cultures in Neurobiology,* Fedoroff, S. and Hertz, L., Eds., Academic Press, New York, 1977, 441.
4. **Hertz, L., Juurlink, B. H. J., Fosmark, H., and Schousboe, A.,** Astrocytes in primary cultures, in *Neuroscience Approached Through Cell Culture,* Vol. 1, Pfeiffer, S. E., Ed., CRC Press, Boca Raton, Fla., 1982, 175.
5. **Young, A. B., Oster-Granite, M. L., Herndon, R. M., and Snyder, S. H.,** Glutamic acid: selective depletion by viral induced granule cell loss in hamster cerebellum, *Brain Res.,* 73, 1, 1974.
6. **Hudson, D. B., Valcana, T., Bean, G., and Timiras, P. A.,** Glutamic acid: a strong candidate as the neurotransmitter of the cerebellar granule cell, *Neurochem. Res.,* 1, 73, 1976.
7. **Stone, T. W.,** Glutamate as the neurotransmitter of cerebellar granule cells in the rat: electrophysiological evidence, *Br. J. Pharmacol.,* 66, 291, 1979.
8. **Hertz, L., Yu, A., Svenneby, G., Kvamme, E., Fosmark, H., and Schousboe, A.,** Absence of preferential glutamine uptake into neurons. An indication of a net transfer of TCA constituents from nerve endings to astrocytes?, *Neurosci. Lett.,* 16, 103, 1980.
9. **Yu, A. C. H. and Hertz, L.,** Uptake of glutamate, GABA, and glutamine into a predominatly GABAergic and a predominantly glutamatergic nerve cell population in culture, *J. Neurosci. Res.,* 7, 23, 1982.
10. **Drejer, J., Larsson, O. M., and Schousboe, A.,** Characterization of L-glutamate uptake into and release from astrocytes and neurons cultured from different brain regions, *Exp. Brain Res.,* 47, 259, 1982.
11. **Drejer, J., Larsson, O. M., and Schousboe, A.,** Characterization of uptake and release processes for D- and L-aspartate in primary cultures of astrocytes and cerebellar granule cells, *Neurochem. Res.,* 8, 231, 1983.
12. **Drejer, J., Larsson, O. M., Kvamme, E., Svenneby, G., Hertz, L., and Schousboe, A.,** Ontogenetic development of glutamate metabolizing enzymes in cultured cerebellar granule cells and in cerebellum, *Neurochem. Res.,* 10, 49, 1985.
13. **Meier, E., Drejer, J., and Schousboe, A.,** GABA induces functionally active low-affinity GABA receptors on cultured cerebellar granule cells, *J. Neurochem.,* 43, 1737, 1984.
14. **Gallo, V., Ciotti, M. T., Coletti, A., Aloisi, F., and Levi, G.,** Selective release of glutamate from cerebellar granule cells differentiating in culture, *Proc. Natl. Acad. Sci. U.S.A.,* 79, 7919, 1982.
15. **Meier, E., Hansen, G. H., and Schousboe, A.,** The trophic effect of GABA on cerebellar granule cells is mediated by GABA-receptors, *Int. J. Dev. Neurosci.,* 3, 401, 1985.
16. **Hansen, G. H., Meier, E., and Schousboe, A.,** GABA influences the ultrastructure composition of cerebellar granule cells during development in culture, *Int. J. Dev. Neurosci.,* 2, 247, 1984.
17. **Levi, G., Aloisi, F., Ciotti, M. T., and Gallo, V.,** Autoradiographic localization and depolarization-induced release of acidic amino acids in differentiating cerebellar granule cells, *Brain Res.,* 290, 77, 1984.
18. **Schousboe, A., Drejer, J., Hansen, G. H., and Meier, E.,** Cultured neurons as model systems for biochemical and pharmacological studies on receptors for neurotransmitter amino acids, *Dev. Neurosci.,* 7, 252, 1986.
19. **Schousboe, A. and Hertz, L.,** Primary cultures of GABAergic and glutamatergic neurons as model systems to study neurotransmitter function. II. Developmental aspects, in *Model Systems of Development and Aging of the Nervous System,* Vernadakis, A. and Giacobini, E., Eds., M. Nijhoff, Boston, 1987, 33.
20. **Ribak, C. E.,** Aspinous and sparsely-spinous stellate neurons in the visual cortex of rats contain glutamic acid decarboxylase, *J. Neurocytol.,* 7, 461, 1978.
21. **Dichter, M. A.,** Rat cortical-neurons in cell culture-culture methods, cell morphology, electrophysiology and synapse formation, *Brain Res.,* 149, 279, 1976.

22. **Dichter, M. A.,** Physiological identification of GABA as the inhibitory transmitter for mammalian cortical neurons in cell culture, *Brain Res.*, 190, 111, 1980.
23. **Hauser, K., Balcar, V. J., and Bernasconi, R.,** Development of GABA neurons in dissociated cell culture of rat cerebral cortex, in *GABA Neurotransmission. Current Developments in Physiology and Neurochemistry,* Lal, H., Fielding, S., Malick, J., Roberts, E., Shah, N., and Usdin, E., Eds., *Brain Res. Bull.*, 5(Suppl. 2), 37, 1980.
24. **Snodgrass, S. R., White, W. F., Biales, B., and Dichter, M.,** Biochemical correlates of GABA function in rat cortical neurons in culture, *Brain Res.*, 190, 123, 1980.
25. **Larsson, O. M., Thorbek, P., Krogsgaard-Larsen, P., and Schousboe, A.,** Effect of homo-β-proline and other heterocyclic GABA analogues on GABA uptake in neurons and astroglial cells and on GABA receptor binding, *J. Neurochem.*, 37, 1509, 1981.
26. **Larsson, O. M., Drejer, J., Hertz, L., and Schousboe, A.,** Ion dependency of uptake and release of GABA and (RS)-nipecotic acid studied in cultured mouse brain cortex neurons, *J. Neurosci. Res.*, 9, 291, 1983.
27. **Yu, A. C. H., Hertz, E., and Hertz, L.,** Alternations in uptake and release rates for GABA, glutamate and glutamine during biochemical maturation of highly purified cultures of cerebral cortical neurons, a GABAergic preparation, *J. Neurochem.*, 42, 951, 1984.
28. **Larsson, O. M., Drejer, J., Kvamme, E., Svenneby, G., Hertz, L., and Schousboe, A.,** Ontogenetic development of glutamate and GABA metabolizing enzymes in cultured cerebral cortex interneurons and in cerebral cortex *in vivo, Int. J. Dev. Neurosci.*, 3, 177, 1985.
29. **Hertz, L.,** Astrocytes, in *Handbook of Neurochemistry,* Vol. 1, 2nd ed., Lajtha, A., Ed., Plenum Press, New York, 1982, 319.
30. **Henn, F. A.,** Neurotransmission and glial cells: a functional relationship?, *J. Neurosci. Res.*, 2, 271, 1976.
31. **Schousboe, A. and Divac, I.,** Differences in glutamate uptake in astrocytes cultured from different brain regions, *Brain Res.*, 177, 407, 1979.
32. **Schousboe, A., Drejer, J., and Divac, I.,** Regional heterogeneity in astroglial cells, *Trends Neurosci.*, 3, 11, 1980.
33. **Drejer, J., Meier, E., and Schousboe, A.,** Novel neuron-related regulatory mechanisms for astrocytic glutamate and GABA high affinity uptake, *Neurosci. Lett.*, 37, 301, 1983.
34. **Hansson, E.,** Accumulation of putative, amino acid neurotransmitters, monoamines and D-ala^2-met-enkephalinamide in primary astroglial cultures from various brain areas, visualized by autoradiography, *Brain Res.*, 289, 189, 1983.
35. **Schousboe, A., Nissen, C., Bock, E., Sapirstein, V., Juurlink, B. H. J., and Hertz, L.,** Biochemical development of rodent astrocytes in primary cultures, in *Tissue Culture in Neurobiology,* Giacobini, E., Vernadakis, A. and Shahar, A., Eds., Raven Press, New York, 1980, 397.
36. **Drejer, J., Benveniste, H., Diemer, N. H., and Schousboe, A.,** The cellular origin of ischemia-induced glutamate release from brain tissue *in vivo* and *in vitro, J. Neurochem.*, 45, 145, 1985.
37. **Schousboe, A., Larsson, O. M., Drejer, J., Krogsgaard-Larsen, P., and Hertz, L.,** Uptake and release processes for glutamine, glutamate and GABA in cultured neurons and astrocytes, in *Glutamine, Glutamate and GABA in the Central Nervous System,* Hertz, L., Kvamme, E., McGeer, E. G., and Schousboe, A., Eds., Alan R. Liss, New York, 1983, 297.
38. **Levi, G. and Ciotti, M. T.,** Glutamate and GABA localization and evoked release in cerebellar cells differentiating in culture, in *Glutamine, Glutamate and GABA in the Central Nervous System,* Hertz, L., Kvamme, E., McGeer, E. G., and Schousboe, A., Eds., Alan R. Liss, New York, 1983, 493.
39. **Hertz, L., Schousboe, A., Boechler, N., Mukerji, S., and Fedoroff, S.,** Kinetic characteristics of the glutamate uptake into normal astrocytes in culture, *Neurochem. Res.*, 3, 1, 1978.
40. **Schousboe, A., Hertz, L., Svenneby, G., and Kvamme, E.,** Phosphate activated glutaminase activity and glutamine uptake in primary cultures of astrocytes, *J. Neurochem.*, 32, 943, 1979.
41. **Drejer, J., Honore, T., Meier, E., and Schousboe, A.,** Pharmacologically distinct glutamate receptors on cerebellar granule cells, *Life Sci.*, 38, 2077, 1986.
42. **Greenamyre, J. T., Young, A. B., and Penney, J. B.,** Quantitative autoradiographic distribution of [^{3}H]glutamate binding sites in rat central nervous system, *J. Neurosci.*, 4, 2133, 1984.
43. **Meier, E., Drejer, J., and Schousboe, A.,** Trophic actions of GABA on the development of physiologically active GABA receptors, in *CNS-Receptors from Molecular Pharmacology to Behavior,* Mandel, P. and DeFeudis, F. V., Eds., Raven Press, New York, 1983, 47.
44. **Meier, E., Drejer, J., and Schousboe, A.,** Inhibition by GABA of evoked glutamate release from cultured cerebellar granule cells, in *Glutamine, Glutamate and GABA in the Central Nervous System,* Hertz, L., Kvamme, E., McGeer, E. G., and Schousboe, A., Eds., Alan R. Liss, New York, 1983, 509.
45. **Belhage, B., Meier, E., and Schousboe, A.,** GABA-agonists induce the formation of low affinity GABA-receptors on cultured cerebellar granule cells via preexisting high affinity GABA-receptors, *Neurochem. Res.*, 11, 599, 1986.

46. **Hill, D. R. and Bowery, N. G.,** ^{3}H-Baclofen and ^{3}H-GABA bind to bicuculline-insensitive $GABA_B$ sites in rat brain, *Nature (London)*, 290, 149, 1981.
47. **Hertz, L.,** Potassium effects on transport of amino acids, inorganic ions and water: ontogenetic and quantitative differences, in *Transport Phenomena in the Nervous System. Physiological and Pathological Aspects*, Adv. Exp. Med. Biol. 69, Levi, G., Battistin, L., and Lajtha, A., Eds., Plenum Press, New York, 1976, 371.
48. **Ramaharobandro, N., Borg, J., Mandel, P., and Mark, J.,** Glutamine and glutamate transport in cultured neuronal and glial cells, *Brain Res.*, 244, 113, 1982.
49. **Borg, J., Spitz, B., Hamel, G., and Mark, J.,** Selective culture of neurons from rat cerebral cortex: morphological characterization, glutamate uptake and related enzymes during maturation in various culture media, *Brain Res.*, 350, 37, 1985.
50. **Hertz, L., Bock, E., and Schousboe, A.,** GFA content, glutamate uptake and activity of glutamate metabolizing enzymes in differentiating mouse astrocytes in primary cultures, *Dev. Neurosci.*, 1, 226, 1978.
51. **Balcar, V. J. and Hauser, K. L.,** Transport of [^{3}H]L-glutamate and [^{3}H]L-glutamine by dissociated glial and neuronal cells in primary culture, *Proc. Eur. Soc. Neurochem.*, 1, 498, 1978.
52. **Lehmann, A., Isacsson, H., and Hamberger, A.,** Effects of in vivo administration of kainic acid on the extracellular amino acid pool in the rabbit hippocampus, *J. Neurochem.*, 40, 1314, 1983.
53. **Benveniste, H., Drejer, J., Schousboe, A., and Diemer, N. H.,** Elevation of the extracellular concentrations of glutamate and aspartate in rat hippocampus during transient cerebral ischemia monitored by intracerebral microdialysis, *J. Neurochem.*, 43, 1369, 1984.
54. **Martin, D. L.,** Carrier-mediated transport and removal of GABA from synaptic regions, in *GABA in Nervous System Function*, Roberts, E., Chase, T. N., and Tower, D. B., Eds., Raven Press, New York, 1976, 347.
55. **Balcar, V. J. and Johnston, G. A. R.,** The structural specificity of the high affinity uptake of L-glutamate and L-aspartate by rat brain slices, *J. Neurochem.*, 19, 2657, 1972.
56. **Roberts, P. J. and Watkins, J. C.,** Structural requirements for the inhibition of L-glutamate uptake by glia and nerve endings, *Brain Res.*, 85, 120, 1975.
57. **Balcar, V. J., Borg, J., and Mandel, P.,** High affinity uptake of L-glutamate and L-aspartate by glial cells, *J. Neurochem.*, 28, 87, 1977.
58. **Balcar, V. J., Johnston, G. A. R., and Twitchin, B.,** Stereospecific inhibition of L-glutamate and L-aspartate high affinity uptake in rat brain slices by threo-3-hydroxyaspartate, *J. Neurochem.*, 28, 1145, 1977.
59. **Schousboe, A., Svenneby, G., and Hertz, L.,** Uptake and metabolism of glutamate in astrocytes cultured from dissociated mouse brain hemispheres, *J. Neurochem.*, 29, 999, 1977.
60. **Drejer, J.,** Glutamate-Mediated Neurotransmission. A Review. Studies in Neurons and Astrocytes in Primary Cultures, Ph.D. thesis, Royal Danish School of Pharmacy, Copenhagen, 1982, 1.
61. **Schousboe, A., Drejer, J., Meier, E., Larsson, O. M., and Schousboe, I.,** Regional heterogeneity of astrocytic transport processes for glutamate and GABA, in *Dynamic Properties of Glial Cells. II. Cellular and Molecular Aspects*, Grisar, T., Franck, G., Hertz, L., Norton, W., Sensenbrenner, M., and Woodbury, D. M., Eds., Pergamon Press, Oxford, 1986, 225.
62. **Schousboe, A., Drejer, J., Larsson, O. M., and Meier, E.,** Regulation of astrocytic high affinity uptake of transmitter amino acids by neuronal signaling, in *Glial-Neuronal Communication in Development and Regeneration. NATO ARW*, Althaus, H. and Seifert, W., Eds., Plenum Press, New York, 1987, 231.
63. **Yu, A. C., Schousboe, A., and Hertz, L.,** Metabolic fate of ^{14}C labeled glutamate in astrocytes in primary cultures, *J. Neurochem.*, 39, 954, 1982.
64. **Yu, A. C. H., Schousboe, A., and Hertz, L.,** Influences of pathological concentrations of ammonia on metabolic fate of ^{14}C-labelled glutamate in astrocytes in primary cultures, *J. Neurochem.*, 42, 594, 1984.
65. **Waniewski, R. A. and Martin, D. A.,** Characterization of L-glutamic acid transport by glioma cells in culture: evidence for sodium-independent, chloride-dependent high affinity influx, *J. Neurosci.*, 4, 2237, 1984.
66. **Schousboe, A., Drejer, J., and Hertz, L.,** Comparison of the metabolism of glucose and glutamate in cultured cerebellar granule cells, *J. Neurochem., Suppl.*, 44, S168, 1985.
67. **Yu, A. C. H. and Hertz, L.,** Metabolic sources of energy in astrocytes, in *Glutamine, Glutamate, and GABA in the Central Nervous System*, Hertz, L., Kvamme, E., McGeer, E. G., and Schousboe, A., Eds., Alan R. Liss, New York, 1983, 431.
68. **Shapiro, R. A., Clark, V. M., and Curthoys, N. P.,** Inactivation of rat renal phosphate-dependent glutaminase with 6-diazo-5-oxo-L-norleucine — evidence for interaction of the glutamine binding site, *J. Biol. Chem.*, 254, 2835, 1979.
69. **Lisy, V., Stastny, F., Murphy, S., and Hajkova, B.,** Glutamate uptake into cerebral cortex slices is reduced in the presence of a gamma-glutamyl transpeptidase inhibitor, *Experientia*, 39, 111, 1983.

70. **Minn, A. and Besagni, D.,** Uptake of L-glutamine into synaptosomes, *Life Sci.*, 33, 225, 1983.
71. **Shine, H. D., Hertz, L., De Vellis, J., and Haber, B.,** A fluorimetric assay for gamma-glutamyl transpeptidase: demonstration of enzymatic activity in cultured cells of neural origin, *Neurochem. Res.*, 6, 453, 1981.
72. **Kvamme, E., Schousboe, A., Hertz, L., Torgner, I. A., and Svenneby, G.,** Developmental change of endogenous glutamate and gamma-glutamyl transferase in cultured cerebral cortical interneurons and cerebellar granule cells and in mouse cerebral cortex and cerebellum *in vivo*, *Neurochem. Res.*, 10, 993, 1985.
73. **McLennan, H.,** The autoradiographic localization of L-[^{3}H]glutamate in rat brain tissue, *Brain Res.*, 115, 139, 1976.
74. **Hertz, L.,** Functional interactions between neurons and astrocytes. I. Turnover and metabolism of putative amino acid transmitters, *Prog. Neurobiol.*, 13, 277, 1979.
75. **Drejer, J. and Schousboe, A.,** unpublished experiments.

Chapter 17

METABOLISM OF GLUTAMATE AND GLUTAMINE IN NEURONS AND ASTROCYTES IN PRIMARY CULTURES

Leif Hertz and Arne Schousboe

TABLE OF CONTENTS

I. INTRODUCTION

In Chapter 16, it was discussed that glutamate is released in a transmitter-related manner from glutamatergic neurons, e.g., cultured cerebellar granule cells, and that to a large extent it is accumulated into astrocytes. This obviously raises two questions: (1) what happens to this glutamate after its astrocytic accumulation and (2) how is neuronal glutamate replenished? In GABAergic neurons, e.g., cultured cerebral cortical neurons, glutamate is decarboxylated to GABA which also is accumulated partly into astrocytes. This will create a similar problem, but quantitatively to a lesser extent, since the amount of released GABA is smaller than that of glutamate.[1-3] Moreover, GABA is accumulated with considerably higher intensity into neurons than into astrocytes.[1-4]

Thus, it is of interest to investigate metabolic conversion leading to the formation or degradation of glutamate in astrocytes and in different types of neurons which either utilize glutamate directly as a neurotransmitter or utilize glutamate as a precursor for GABA. This chapter will review current knowledge about activities of enzymes pertinent to these metabolic pathways as well as actual metabolic fluxes in neurons and astrocytes in primary culture. Such cultures seem to constitute reliable models for their in vivo counterparts.[5-12] Studies on cell lines will not be included because of the risk of quantitative alterations in metabolic processes in these transformed cells.[11-13] A brief description of the cultured cells is given in Chapter 16.

II. METABOLIC PATHWAYS FOR GLUTAMATE AND GLUTAMINE

A. Pathways

Pathways which potentially could be involved in either conversion of glutamate to other compounds or in formation of glutamate from its precursors are illustrated in Figure 1. Glutamate can be converted to GABA (a process occurring only in GABAergic neurons), glutamine, or α-ketoglutarate (2-oxoglutarate), a tricarboxylic acid (TCA) cycle constituent. The latter process can occur either as a transamination, catalyzed by aspartate aminotransferase (AAT) (Chapter 8, Volume I), or as an oxidative deamination, catalyzed by glutamate dehydrogenase (GLDH) (Chapter 6, Volume I). A distinction between these possibilities can be achieved by the use of aminooxyacetic acid (AOAA), an inhibitor of the transamination but not of the oxidative deamination. The formation of glutamine, which requires ATP, is catalyzed by glutamine synthetase (GS) (Chapter 2, Volume I), and that of GABA is catalyzed by glutamate decarboxylase (GAD) (Chapter 7, Volume I).

The synthesis of glutamate from its precursors (Figure 1) is even more complex than glutamate degradation. Glutamate formation from glutamine cannot proceed as a reversal of the GS reaction and is catalyzed by a different enzyme, i.e., phosphate-activated glutaminase, PAG (Chapter 3, Volume I). This makes it possible to obtain an estimate of the relative importance of the two directions of the reaction by measuring activities of GS and PAG. Glutamate can also be derived from α-ketoglutarate, either by a transamination with aspartate catalyzed by AAT (Figure 1) or with a branched-chain amino acid (Section III.D.6) or by a reductive amination catalyzed by GLDH. Both of these enzymes catalyze the reactions in either direction, so that measurement of AAT or GLDH activity by itself gives no information whether the reaction occurs in one or the other direction. Since α-ketoglutarate

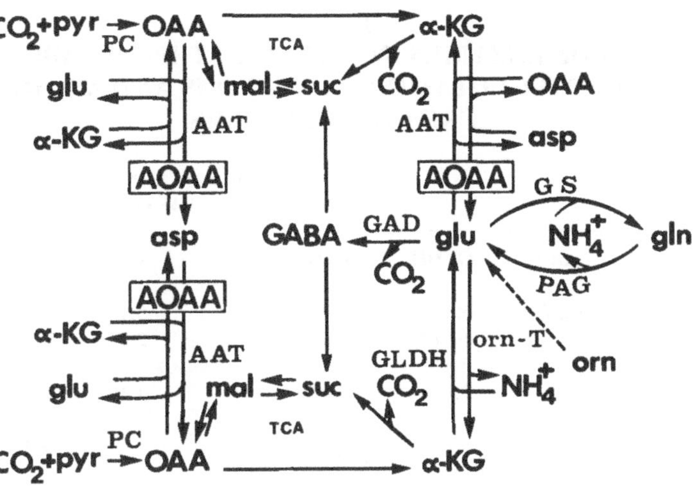

FIGURE 1. Scheme of metabolic interconversions between glutamate (glu), glutamine (gln), γ-aminobutyric acid (GABA), ornithine (orn), and the tricarboxylic acid (TCA) cycle constituents, succinate (suc), malate (mal), oxaloacetate (OAA), and α-ketoglutarate (α-KG); pyruvate (pyr), aspartate (asp), carbon dioxide (CO_2), and ammonia (NH_4^+). Some of the enzymes catalyzing these reactions are indicated: glutamine synthetase (GS), phosphate-activated glutaminase (PAG), glutamate dehydrogenase (GLDH), aspartate aminotransferase (AAT), glutamate decarboxylase (GAD), ornithine aminotransferase (orn-T), and pyruvate carboxylase (PC). The reactions inhibited by aminooxyacetic acid (AOAA) are indicated by ▭. (Modified from Hertz, L., Yu, A. C. H., Potter, R. L., Fisher, T. E., and Schousboe, A., *Glutamine, Glutamate, and GABA in the Central Nervous System*, Hertz, L., Kvamme, E., McGeer, E. G., and Schousboe, A., Eds., Alan R. Liss, New York, 1983, 327.)

is a TCA cycle constitutent, a net synthesis of this compound, and thus of glutamate, requires an anaplerotic process such as pyruvate carboxylation, i.e., a condensation of pyruvate with bicarbonate to form oxaloacetatate.[15] This process is catalyzed by pyruvate carboxylase (PC) at the expense of ATP. Accordingly, the PC activity is also of interest when pathways that can lead to glutamate synthesis are considered. An alternative source for glutamate formation is ornithine which by the concerted action of ornithine-δ-aminotransferase (Orn-T) and glutamate semialdehyde dehydrogenase or the Δ^1-pyrrolidine-5-carboxylate dehydrogenase can be converted into glutamate.[16-18] Ornithine may be derived from arginine by the action of arginase.[16]

One more enzyme which may be of relevance in this context is the γ-glutamyltransferase (γ-glutamyltranspeptidase), γ-GT, which has been suggested to play a role in membrane transport of amino acids[19] such as glutamine or glutamate uptake in synaptosomes[20] and brain slices.[21] A discussion of these aspects can be found in Chapter 16.

B. Enzyme Activities in Neurons

1. Cerebellar Granule Cells

Specific activities of enzymes involved in glutamate biosynthesis or glutamate degradation are shown in Table 1. The activity of AAT is exceedingly high which is in agreement with suggestions by Wenthold and Altschuler[40] that this enzyme is primarily associated with glutamatergic neurons. The activities of GLDH and PAG are also relatively high, although they are at least one order of magnitude lower than AAT activity. In contrast, GS activity is low and it is unresolved whether the activity of this enzyme is higher than can be explained by a moderate contamination of the cultures with astrocytes. Orn-T activity is also low and comparable to the γ-GT activity. The phosphate activation of PAG is more pronounced in

Table 1
ACTIVITIES OF ENZYMES INVOLVED IN THE METABOLISM OF GLUTAMATE AND GLUTAMINE IN NEURONS AND ASTROCYTES IN PRIMARY CULTURES

	Enzyme activities[a] (nmol × min^{-1} × mg^{-1})							
Cell type	**AAT**	**GLDH**	**PAG**	**GAD**	**GS**	**Orn-T**	**PC**	**γ-GT**
Cerebellar granule cells	1657[9]	60.8[b22]	21.8[c22]		5.7[22]	3.1[23]	0[24]	2.2[10]
		62.8[d9]	5.6[e9]		2.7[9]	3.1[9]		
		12.5[9]	19.0[f9]					
Cerebral cortical neurons	280[8]	25.4[8]	7.0[e8]	1.3[25]	4.1[8]	5.3[23]	0[24]	1.1[10]
			18.8[f8]		7.9[26]			
			3.9[g26]	10[27]		5.3[8]		
				14.2[7]				
				6.0[26]				
Astrocytes	206[28]	12.3[28]	3.8[e29]	0.009[30]	25.9[28]	5.3[23]	1.8[24]	1.0—2.2[h31]
	154—249[5]	16.6[32]	10.7[f29]	<0.05[28]	22.0—28.1[5]			
			2.4[e33]					
	120—140[34]	3.0—12.3[5]	8.9[f33]					
			1.8[c2]	0.2[35]	41.3[i36]			
		10[34]	1.5[26]		43.3[i37]			
		101.3[b]			10[34]			
					22.2[22]			
					90.3[26]			
					76.7[38]			
					11.3[39]			

[a] For definition of abbreviations, see Section II.
[b] Assayed in the presence of ADP.
[c] This value has been multiplied by 10 since the authors indicated that only 10% of maximum activity occurred at the glutamine concentration used.
[d] 8-day-old cultures.
[e] Assayed at 5 m*M* phosphate.
[f] Assayed at 20 m*M* phosphate.
[g] Assayed at 150 m*M* phosphate.
[h] The reported value has been corrected from μmol to nmol.
[i] Treated with glucocorticoids.

the cerebellar granule cell cultures than in the cerebellum in vivo.[9,41] GAD activity has not been determined in cultured cerebellar granule cells since from histochemical investigations in vivo it is known that GAD is absent from these cells.[42] Also PC has been found to be absent from cerebellar granule cells (Table 1).

The development of these enzymes during the culture period has been followed in some detail.[9,10,41] With the exception of GS, all enzyme activities develop *pari passu* with their development in the cerebellum in vivo. If anything, the development is accelerated compared to the in vivo situation.[9,41] This is of practical importance since the cultured cells can only be maintained for 2 to 3 weeks. A remarkable feature both in vivo and in cultured granule cells is that GLDH activity reaches a peak which coincides with the development of processes. An analogous peak of GLDH activity during process formation has been reported for hippocampus.[43]

2. Cerebral Cortical Neurons

In contrast to cerebellar granule cells, the cerebral cortical neurons have a high GAD activity (Table 1). PAG activity is similar to that in cerebellar granule cells, whereas both

AAT and GLDH activities are somewhat lower, although they still are high compared to the activities of the other enzymes. Orn-T activity in the cerebral neurons is comparable to that in cerebellar granule cells, but a striking difference between this enzyme in the two cell types is that the K_m for ornithine is much lower in cerebral cortical neurons (0.5 m*M*) than in cerebellar granule cells (5 m*M*).[23] As in cerebellar granule cells, no PC activity can be found (Table 1). As was the case with cerebellar granule cells, the development of these enzymes, including GAD, is somewhat faster than in the cerebral cortex in vivo and the activities reach the same or higher (five times in the case of GAD) levels.[7,8,41]

C. Enzyme Activities in Astrocytes

A substantial amount of information, originating from different laboratories, about enzyme activities in astrocytes is now available (Table 1). In general the agreement between different laboratories is good. Thus, a high GS activity has consistently been found, whereas the activity of GAD is negligible. This is in agreement with the immunohistochemical demonstration that GS is an astrocytic enzyme[44] and that GAD is a specific marker for GABAergic neurons (Chapter 7, Volume I). The activity of GS is enhanced after exposure to glucocorticoids[6,36-39,45] and the induced activity is twice as high as GS activity in vivo. AAT activity is comparable to that in cerebral cortical neurons, whereas both GLDH and PAG activities are slightly lower than in neurons. Nevertheless, the activities of both enzymes are substantial. Orn-T activity is comparable to that in neurons, but the K_m is as high as in the cerebellar granule cells. Another difference from the two types of neurons is that there is a substantial PC activity (1 to 2 nmol $\times$ min^{-1} $\times$ mg^{-1}).[24] This is in agreement with histochemical data that have shown this enzyme to be present in glial cells, but absent in neurons in the cerebellum in vivo.[46,47] The development of GS is similar in cultured astrocytes and in the cerebral cortex in vivo.[5,34,38] AAT and GLDH activities develop more slowly than in neurons in primary cultures and follow the development in vivo more closely.[6]

III. METABOLIC FLUXES OF GLUTAMATE AND GLUTAMINE

A. Oxidative Metabolism

1. Neurons

a. Cerebellar Granule Cells

The rate of CO_2 production from [1-^{14}C]glutamate, which reflects the rate of decarboxylation of α-ketoglutarate formed from glutamate, is negligible in cerebellar granule cells (Table 2). This is the case regardless of whether or not AOAA is present.[48] This is not only an indication that these cells use very little glutamate as a metabolic substrate, but also that the metabolism of glutamate is not invariably high in cultured cells.

b. Cerebral Cortical Neurons

The rate of formation of CO_2 from [1-^{14}C]glutamate seems to be somewhat higher in cerebral cortical neurons that in cerebellar granule cells (Table 2). In the absence of AOAA it will, however, include decarboxylation of glutamate to GABA, and in the presence of AOAA, which inhibits both GABA production and transamination of glutamate, it is about 0.4 nmol $\times$ min^{-1} $\times$ mg^{-1}. The CO_2 production rate from [1-^{14}C]glutamine (in the presence of AOAA) is considerably higher (Table 2) and exceeds the oxidation rate for glucose.[53,54] In the presence of an elevated ammonia concentration (3 m*M*) the oxidative metabolism of glutamate is decreased by about 60%.[55]

2. Astrocytes

Experiments in the authors' laboratories have consistently demonstrated a very substantial oxidative degradation of both [1-^{14}C]- (Table 2) and [U-^{14}C]-[52] glutamate in astrocytes. In

Table 2
RATES OF CARBON DIOXIDE PRODUCTION IN NEURONS AND ASTROCYTES IN PRIMARY CULTURES AFTER INCUBATION WITH GLUTAMINE AND GLUTAMATE

Cell type	Substrate	CO_2 (nmol × min^{-1} × mg^{-1})
Cerebellar granule cells	[1-^{14}C]glutamate	0.2[48]
Cerebral cortical neurons	[1-^{14}C]glutamate	0.7[a]
		0.5[b]
		0.4[a,c]
	[1-^{14}C]glutamine	1.8[c,d49]
Astrocytes	[1-^{14}C]glutamate	4.1[d50]
		5.9[d51]
		7.9[d52]
		1.4[b]
		1.6[d,e]
	[1-^{14}C]glutamine	2.0[d]

[a] Hertz, L., Yu, A. C. H., and Schousboe, A., unpublished experiments.
[b] Lai, J. C. K., Murthy, Ch. R. K., Cooper, A. J. L., and Hertz, L., unpublished experiments.
[c] Measured in the presence of 5 m*M* AOAA.
[d] Calculated from specific activity in the tissue.
[e] Hertz, L. and Shargool, M., unpublished experiments.

contrast, Martin et al.[56] have reported a considerably slower rate of carbon dioxide formation. The rate of CO_2 formation from glutamate is little affected in the presence of 5 m*M* AOAA,[50,51] indicating that it occurs as an oxidative deamination (which is resistant to AOAA, Figure 1) and not as a transamination. The fluxes are about one half of GLDH activities (Table 1). As in cerebral cortical neurons, the glutamate oxidation is inhibited by elevated ammonia concentrations,[51,55,57] although to a lesser extent than in neurons.

The conclusion that the GLDH-catalyzed reaction proceeds toward production of carbon dioxide and not toward the synthesis of glutamate may appear to be at variance with the fact that the equilibrium constant of the GLDH-catalyzed reaction favors glutamate formation. However, it can be calculated[50] that with the concentrations of α-ketoglutarate, ammonia, NADH, and NAD^+ presumably present in the mitochondria this reaction would be in equilibrium at an intracellular glutamate concentration of 80 nmol × mg^{-1}, i.e., less than the actual glutamate content in astrocytes which amounts to about 100 to 125 nmol × mg^{-1}.[50,51] The reason why the reaction in astrocytes proceeds toward α-ketoglutarate formation seems thus to be the very high glutamate content in the cells.

B. Formation of GABA

1. Neurons

a. Cerebellar Granule Cells

No GABA formation can be demonstrated after labeling with glutamine.[87] Moreover, the carbon dioxide formation from [1-^{14}C]glutamate (Table 2) is neither inhibited by AOAA (Section III.A.1.a) nor by 3-mercaptopropionic acid or allylglycine, two relatively specific GAD inhibitors.[58]

b. Cerebral Cortical Neurons

It is in agreement with the presence of GAD and glutaminase at high activities in cerebral cortical neurons that glutamine is converted to GABA in these cells. The rate of formation

Table 3
RATES OF GLUTAMINE AND GABA FORMATION IN NEURONS AND ASTROCYTES IN PRIMARY CULTURES

		Rate of synthesis ($nmol \times min^{-1} \times mg^{-1}$)	
Cell type	**Substrate**	**Glutamine**	**GABA**
Cerebral cortical neurons	^{14}C-glutamine		0.3[49]
Astrocytes	^{14}C-glutamate	2.4[50]	
		1.7[51]	
		1.2[59]	
	^{14}C-malate	1.3[60]	

is approximately 0.3 nmol $\times$ min^{-1} $\times$ mg^{-1} (Table 3) which is considerably less than the activity of either enzyme (Table 1). It is in agreement with the concept of formation of GABA from glutamine that there is a potassium-induced release of radioactivity from cerebral cortical neurons after labeling with [U-^{14}C]glutamine.[7,61] In contrast, there is no potassium-induced release of radioactivity after labeling with glutamate[7,27] in spite of the fact that glutamate is accumulated into these cells (Chapter 16). This seems to suggest heterogeneity of glutamate pools in the cells, i.e., that there is a glutamate pool which is accessible to exogenous glutamine (but not to exogenous glutamate) and that glutamate in this pool serves as a precursor for GABA.

2. Astrocytes

It is in agreement with the negligible GAD activity in astrocytes (Table 1) that no GABA is formed from labeled glutamate[62] or glutamine.[88]

C. Formation of Glutamine

1. Neurons

a. Cerebellar Granule Cells

In accordance with the low GS activity in granule cells (Table 1) and the notion that GS is an astrocyte specific enzyme,[44] no radioactivity could be observed in glutamine in granule cells after exposure to ^{14}C-labeled malate, although the glutamate pool showed a substantial incorporation of radioactivity.[89] For a more detailed discussion of malate metabolism, see Section III.D.3.a.

b. Cerebral Cortical Neurons

It has been reported by Borg et al.[63] that cultures of cerebral cortical neurons may incorporate label from [^{14}C]-glutamate to glutamine. However, the extent to which this occurs is modest, probably reflecting the low GS activity in these cultures (Table 1). Whether this process occurs in the neurons or in contaminating astroyctes in the cultures is presently unresolved.

2. Astrocytes

Formation of glutamine from glutamate in astrocytes has repeatedly been demonstrated.[50,51,59,62] The procedure used by the present authors for calculation of fluxes is illustrated in Figure 2. Cultures of astrocytes were incubated with [U-^{14}C]-labeled glutamate and the course of alterations in specific activity in glutamate and related amino acids (glutamine and aspartate) was followed. Since the specific radioactivity of glutamate became almost constant after 5 min and that of glutamine leveled off after 15 min, the rates of glutamine synthesis were calculated during the period of 5 to 8 min, i.e., after virtual equilibrium of

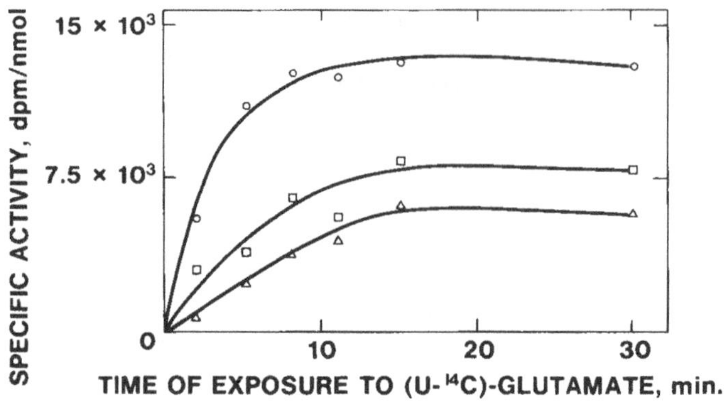

FIGURE 2. Specific radioactivities of glutamate (○), glutamine (△), and aspartate (□) in primary cultures of astrocytes as a function of incubation time with 50 μM L-[U-^{14}C] glutamate. (From Yu, A. C., Schousboe, A., and Hertz, L., *J. Neurochem.*, 39, 954, 1982. With permission.)

the precursor pool and at a time when the glutamine formation rates were still relatively close to initial velocities. During this period the specific radioactivity in glutamine (the product) showed an increase per minute of 420 dpm/nmol. With a pool size of glutamine, which was determined to be 78 nmol/mg protein, this corresponds to an increase of 3.3×10^4 dpm $\times$ min^{-1} $\times$ mg^{-1} of protein in the total radioactivity of the glutamine pool. Since this glutamine was derived from glutamate (the precursor) having a specific radioactivity of 1.4×10^4 dpm $\times$ nmol^{-1} (Figure 2), the amount of glutamine formed per minute must have been $3.3 \times 10^4/1.4 \times 10^4 = 2.4$ nmol/mg of protein. A similar approach, i.e., calculation of fluxes from the increase in specific activity of the product, pool size of the product, and specific activity of the precursor, will be used in the following to determine other flux rates. In our hands, the glutamine formation rate has reproducibly been found to be about 2 nmol $\times$ min^{-1} $\times$ mg^{-1} (Table 3), calculated on the basis of the specific activity of the total glutamate pool in the tissue. A slightly lower (Table 3) maximum value for glutamine synthesis in corresponding rat cultures was obtained in a somewhat different fashion by Waniewski and Martin.[59] These results clearly indicate that glutamate in astrocytes can be converted into glutamine, but the formation rate is only about one half of that with which we have found glutamate to be oxidized (Section II.A.2). Glutamine in astrocytes can also be formed from such glutamate precursors as malate (Section III.D.3.c). This procedure allows a distinction between rates of glutamine synthesis from exogenous and endogenous glutamate. If anything, the rate of glutamine synthesis from the endogenous pool (i.e., from malate-derived glutamate) was lower than that of glutamine synthesis from exogenous glutamate (Table 3).

A prerequisite for glutamine synthesis is a supply of ammonia. Yudkoff et al.[64,65] have determined fluxes from ^{15}N ammonia into glutamine and glutamate in primary cultures of astrocytes (Figure 3). This figure shows the rate of incorporation of ammonia ^{15}N into glutamine and glutamate indicated as isotope enrichment. Such a graph can be treated in the same way as graphs showing specific activities, i.e., by determining increase per time unit in isotope enrichment in the product (glutamine or glutamate), pool size of the product, and isotope enrichment of the precursor (ammonia). From an increase in isotope enrichment in glutamine of 0.4%/min, a glutamine content of 220 nmol $\times$ mg^{-1}, and an isotope enrichment of 20% in tissue ammonia, a rate of 8.8 nmol $\times$ min^{-1} $\times$ mg^{-1} can be calculated for ammonia incorporation into glutamine. This rate is higher than the rate of synthesis of glutamine calculated from incorporation of radioactivity from either ^{14}C-labeled malate or ^{14}C-labeled glutamate (see above). However, it represents a maximum rate calculated on

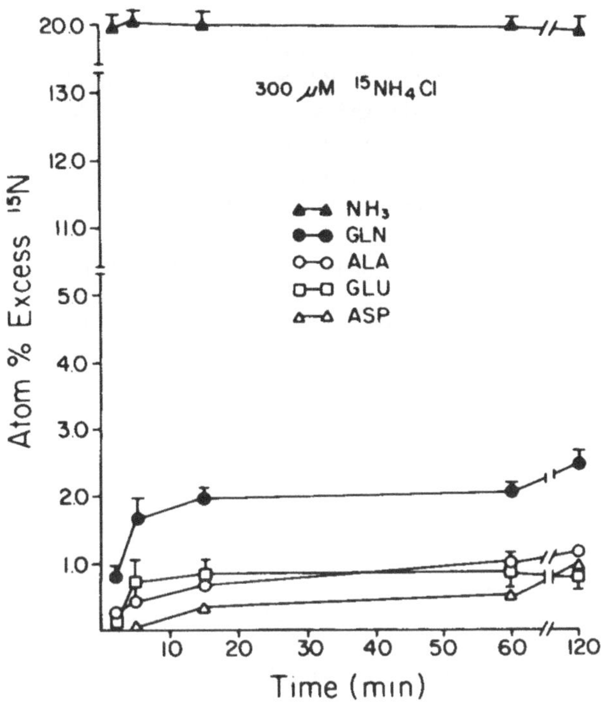

FIGURE 3. ^{15}N enrichment (atom percent excess ^{15}N) in ammonia (▲), glutamine amide N (●), glutamate and glutamine amino N (□), alanine (○), and aspartate (△) as a function of incubation time with 300 μM ^{15}N ammonium chloride. (From Yudkoff, M., Nissim, I., Pleasure, D., Kim, S., Hummeler, K., and Segal, S., *Glutamine, Glutamate and GABA in the Central Nervous System*, Hertz, L., Kvamme, E., McGeer, E. G., and Schousboe, A., Eds., Alan R. Liss, New York, 1983, 389. With permission.)

the assumption that the total ammonia pool in the tissue is available for glutamine synthesis. If this is not the case, the enrichment in the precursor pool would be higher (to a maximum of 100%) and the incorporation rate would accordingly be lower (up to five times).

D. Glutamate Precursors

1. General Considerations

Since a net transport occurs of glutamate from neurons to astrocytes (Chapter 16), it is of interest to know how neuronal glutamate pools are replenished. Moreover, glutamate serves at least two different functions in neurons, i.e., a transmitter-related function and a function as a metabolic substrate which shows a rapid turnover. The purpose of this section is therefore to discuss the role of different potential glutamate precursors both for the formation of cellular pools of glutamate and of transmitter-related glutamate, i.e., glutamate which is released in a potassium-stimulated, calcium-dependent manner.

2. Glutamine

a. Cerebellar Granule Cells

It has long been known that glutamine may serve as a precursor for releasable glutamate in vivo and in vitro.[66,67] A semiquantitative estimate of incorporation of radioactivity from glutamine and other glutamate precursors into cellular and releasable glutamate pools in cerebellar granule cells is shown in Table 4. It can be seen that glutamine labels the cellular

Table 4
LABELING OF CELLULAR AND RELEASABLE POOLS OF GLUTAMATE (GLU) AND ASPARATE (ASP) FROM DIFFERENT PRECURSORS IN CEREBELLAR GRANULE CELLS IN PRIMARY CULTURES[a]

	Cellular pool		Releasable pool[b]	
Precursor	Glu	Asp	Glu	Asp
Glutamine	+ +	0	+ +	0
Malate	+	+ + +	+	(+)
Glucose	+	+ +	(+)	+

Note: 0 = no labeling, + = slight labeling, + + = moderate labeling, and + + + = intense labeling.

[a] Drejer, J., Schousboe, A., Shargool, M., and Hertz, L., unpublished results.
[b] Calcium-dependent, potassium-induced release.

glutamate pool, but not the aspartate pool, and that radioactivity from the labeled glutamate pool is released in a transmitter-related fashion. Formation of aspartate from glutamine via glutamate would have required conversion of glutamate to α-ketoglutarate and subsequent oxidation of this to succinate. The lack of labeling of aspartate is therefore consistent with the fact that very little glutamate is metabolized to carbon dioxide in these cells (Table 2).

b. Cerebral Cortical Neurons

In cerebral cortical neurons glutamine is converted to glutamate at a rate of 1 to 2 nmol $\times$ min^{-1} $\times$ mg^{-1}.[49,54] In our hands,[49] the specific activity of glutamate becomes only about one half of that of glutamine, whereas Borg et al.[63] have observed a higher specific activity in glutamate (the product) than in glutamine (the precursor), suggesting heterogeneity of the glutamine pools in the tissue. In contrast to what was found for cerebellar granule cells, the specific activity of aspartate becomes roughly identical to that in glutamate.[63,68] This is consistent with a higher rate of glutamate oxidation in these cells than in cerebellar granule cells (Table 2). The rate of glutamate formation from glutamine is about one third of the glutaminase activity at a low phosphate concentration (Table 1). The further conversion of glutamine-derived glutamate to GABA was discussed in Section III.B.1.b and amounts to $\simeq$0.3 nmol $\times$ min^{-1} $\times$ mg^{-1} (Table 3). This is only one sixth of the rate with which glutamine undergoes oxidative degradation (Table 2).

c. Astrocytes

In astrocytes, glutamine is also rapidly (2 nmol $\times$ min^{-1} $\times$ mg^{-1}) converted into glutamate.[14,63,69] The rate of conversion is similar to that in cerebral cortical neurons (see above) which is consistent with the similarity between glutaminase activities in the two cell types.

3. Tricarboxylic Acid Cycle Constituents

a. Cerebellar Granule Cells

It has been suggested by Shank and Campbell[70,71] that α-ketoglutarate and malate may be utilized by neurons as precursors for glutamate and/or GABA. Both of these TCA cycle constituents were found to be accumulated into synaptosomes by a high-affinity uptake with

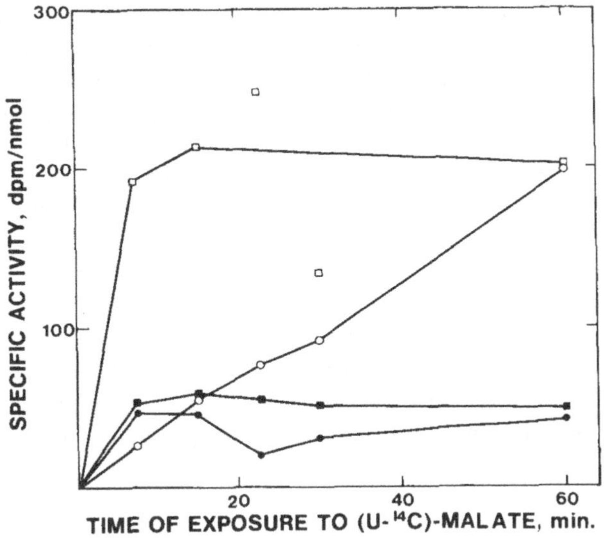

FIGURE 4. Specific radioactivities in the absence (open symbols) or in the presence (closed symbols) of 1 m*M* AOAA of aspartate (□) and glutamate (○) in primary cultures of cerebellar granule cells as a function of incubation time with 10 μ*M* L-U-^{14}C malate. (From Schousboe, A., Drejer, J., Shargool, M., and Hertz, L., unpublished data.)

V_{max} values of about 0.5 nmol × min^{-1} × mg^{-1}.[70] In cerebellar granule cells in primary cultures, a similar, but less intense high-affinity uptake system is found (K_m, 100 μ*M*; V_{max}, 0.1 nmol × min^{-1} × mg^{-1}). After exposure to labeled malate (10 μ*M*), there is a rapid labeling of aspartate and a more gradual labeling of glutamate in the cellular pools (Figure 4). In the presence of AOAA (1 m*M*), the labeling of both aspartate and glutamate is, after a brief latency period, virtually abolished. This suggests either that most, if not all, of the glutamate is formed from malate by a transamination catalyzed by AAT (Figure 1) or that the uptake of malate into the mitochondria might be dependent upon the operation of the malate aspartate shuttle,[72] which is inhibited by AOAA. No rates of glutamate formation can be calculated because the specific activities of malate and its TCA cycle derivatives in the tissue are unknown.

In spite of the more pronounced labeling of the cellular aspartate pool than of the cellular glutamate pool after exposure to labeled malate, there is virtually no evoked release of aspartate, whereas there is a distinct evoked release of glutamate (Table 4). This may either indicate that cerebellar granule cells do not utilize aspartate as their transmitter or that the aspartate pool which is labeled after exposure to malate is not a transmitter-related pool. It is in favor of the latter concept that after labeling from glucose there is an evoked release of labeled aspartate (Table 4).

b. Cerebral Cortical Neurons

Formation of labeled glutamate from [^{14}C]malate has also been demonstrated in cerebral cortical neurons, but, again, the uptake rate is modest.[60] The formation of glutamate appears to be less inhibited by AOAA than in cerebellar granule cells, indicating that it may partly occur as a reductive amination.[88]

c. Astrocytes

In astrocytes, there is a pronounced labeling of glutamate after exposure to ^{14}C-labeled malate.[54,68] A large inhibition of this reaction by AOAA suggests that it mainly occurs as a transamination which is in agreement with the fact that the equilibrium of the GLDH catalyzed process in astrocytes favors the formation of α-ketoglutarate from glutamate, not the reverse reaction (Section II.A.2). The labeled glutamate is further converted to glutamine. A rough estimate can be made of the glutamine synthesis rate based upon the specific activities of the pools of glutamate and glutamine. This rate was calculated to be about 1.3 nmol $\times$ min^{-1} $\times$ mg^{-1}, i.e., slightly lower than the rate of glutamine synthesis from an

4. Ornithine

It has been suggested by Wong et al.,[73] Yoneda and Roberts,[74] and Yoneda et al.[75] that ornithine may serve as a precursor for glutamate. This is in agreement with the metabolic pathways discussed in Section II.A. No information is available about the quantitative role of ornithine as a glutamate precursor in neurons or astrocytes in primary cultures, but the rate with which radioactive ornithine is converted to glutamate in synaptosomes is modest.[74-76]

5. Glucose

a. Neurons

It has often been suggested that glucose serves as a precursor of transmitter amino acids in neurons.[67,77,78] After labeling with [^{14}C]glucose, both cellular pools of aspartate and glutamate and releasable pools of these two amino acids in cerebellar granule cells are labeled (Table 4). This labeling does, however, not mean that a net synthesis of glutamate or aspartate has taken place, since it is a prerequisite for a net conversion of glucose to, for example, glutamate that an anaplerotic process occurs, e.g., a carbon dioxide fixation catalyzed by PC (Figure 1). Since this enzyme is absent from neurons,[24,46,47] no net synthesis of glutamate or aspartate can occur in isolated neurons.

b. Astrocytes

In contrast to neurons, astrocytes do have a sizeable activity of PC (Table 1) so a net synthesis of TCA cycle constituents can occur from glucose in astrocytes and these TCA cycle constituents may be transferred to neurons as discussed above.

6. Branched-Chain Amino Acids

The branched-chain amino acids, leucine, isoleucine, and valine, are also sources of glutamate carbon,[80] but only isoleucine and valine can sustain a net synthesis of the glutamate skeleton in the absence of pyruvate oxidation. The initial steps in the degradation of these amino acids comprise a transamination (yielding nitrogen for glutamate synthesis) and a subsequent decarboxylation. The rate of decarboxylation of leucine has been followed by measurement of $^{14}CO_2$ formation from [1-^{14}C]-labeled leucine and it is considerably higher in astrocytes in primary cultures (about 0.1 nmol $\times$ min^{-1} $\times$ mg^{-1}) than in corresponding cultures of neurons[81] (0.03 nmol $\times$ min^{-1} $\times$ mg^{-1}). Chronic exposure to ammonia causes an increase in this process, probably reflecting the augmentation in leucine aminotransferase activity which has been observed in the brains of rats injected with large doses of ammonium acetate.[83]

IV. CONCLUDING REMARKS

In Chapter 16, it was concluded that uptake and release characteristics for glutamate and glutamine in neurons and astrocytes in primary cultures suggest a large transfer of glutamate from neurons to astrocytes and a considerably smaller return of glutamine to neurons. This indicates that compounds other than glutamine probably also serve as precursors for releasable glutamate and GABA. The present paper supports this conclusion by demonstrating that only a minor part of the glutamate accumulated into astrocytes is converted to glutamine, whereas a major part is oxidized. The validity of this finding is corroborated by the finding that glutamate and glutamine are also converted to CO_2 in brain slices[52] or homogenates and in isolated brain cells.[84] Other cells in culture[85] or in vivo (e.g., the intestinal mucosa)[86] are analogously known to utilize glutamate and/or glutamine as a major substrate for energy metabolism (Chapter 14, Volume I).

The glutamine taken up by cerebral cortical cells was found to be only partly converted to GABA, whereas, again, a considerable part is oxidized. This accentuates the need for alternative glutamate and GABA precursors. The ability of malate to serve as a glutamate precursor was discussed, and it was demonstrated that glutamate production from malate in cerebellar granule cells is virtually abolished in the presence of AOAA. This is a confirmation of the importance of AAT activity in glutamatergic neurons,[40] but provides no information whether glutamate is formed from α-ketoglutarate by a transamination (which would require expenditure of aspartate in glutamate biosynthesis) or whether the uptake of malate into the mitochondria is dependent upon the malate-aspartate shuttle and thus AAT activity. Although the uptake rate for malate into neurons is modest, it would be an attractive concept if glutamine and malate (or any other TCA cycle constituent) were utilized together as glutamate precursors. Utilization of the same amount of glutamine and of a nonamino acid glutamate precursor would ensure that the supply of ammonia equivalents is equal to the amounts required for glutamate and GABA synthesis. In addition, neosynthesis of TCA cycle constituents from glucose can take place via carboxylation of pyruvate to oxaloacetate, and glucose can undoubtedly function as a precursor of transmitter glutamate and GABA. It seems thought provoking that the enzyme catalyzing the formation of oxaloacetate, i.e., the pyruvate carboxylase, has an astrocytic, not a neuronal, localization. This localization is remarkable since it might provide the possibility for a complete astrocytic control over the supply of not only glutamine, but also other precursors for amino acid transmitter synthesis in neurons. Thus, on one hand, neurons may partly regulate astrocytic metabolism by alterations in the supply of glutamate and other substrates, and on the other hand, astrocytes might control the formation of this glutamate. Such a close metabolic interaction between two autonomous cell types might be of fundamental importance for integration of brain function and metabolism. Moreover, disruption of such interactions under pathological conditions, e.g., when the astrocytic glutamate uptake is reduced under ischemia (Chapter 16) or when the oxidative metabolism of glutamate is impaired during exposure to elevated ammonia concentrations, may contribute to the abnormalities in brain function under such conditions.

REFERENCES

1. **Hertz, L.,** Functional interactions between neurons and astrocytes. I. Turnover and metabolism of putative amino acid transmitters, *Proc. Neurobiol.,* 13, 277, 1979.
2. **Schousboe, A.,** Transport and metabolism of glutamate and GABA in neurons and glial cells, *Int. Rev. Neurobiol.,* 22, 1 1981.
3. **Schousboe, A., Larsson, O. M., Drejer, J., Krogsgaard-Larsen, P., and Hertz, L.,** Uptake and release processes for glutamine, glutamate and GABA in cultured neurons and astrocytes, in *Glutamine, Glutamate and GABA in the Central Nervous System,* Hertz, L., Kvamme, E., McGeer, E. G., and Schousboe, A., Eds., Alan R. Liss, New York, 1983, 297.
4. **Hertz, L. and Schousboe, A.,** Primary cultures of GABAergic and glutamatergic neurons as model systems to study neurotransmitter functions. I. Differentiated cells, in *Model Systems of Development and Aging of the Nervous System,* Vernadakis, A., Privat, A., Lauder, J. N., Timiras, P. S., and Giacobini, E., Eds., Kluwer-Nijhoff, Boston, 1986, 19.

5. **Hertz, L., Bock, E., and Schousboe, A.,** GFA content, glutamate uptake and activity of glutamate metabolizing enzymes in differentiating mouse astrocytes in primary cultures, *Dev. Neurosci.*, 1, 226, 1978.
6. **Schousboe, A., Nissen, C., Bock, E., Sapirstein, V. S., Juurlink, B. H. J., and Hertz, L.,** Biochemical development of rodent astrocytes in primary cultures, in *Tissue Culture in Neurobiology*, Giacobini, E., Vernadakis, A., and Shahar, A., Eds., Raven Press, New York, 1980, 397.
7. **Yu, A. C. H., Hertz, E., and Hertz, L.,** Alterations in uptake and release rates for GABA, glutamate and glutamine during biochemical maturation of highly purified cultures of cerebral cortical neurons, a GABAergic preparation, *J. Neurochem.*, 42, 951, 1984.
8. **Larsson, O. M., Drejer, J., Kvamme, E., Svenneby, G., Hertz, L., and Schousboe, A.,** Ontogenetic development of glutamate and GABA metabolizing enzymes in cultured cerebral cortex interneurons and in cerebral cortex *in vivo*, *Int. J. Dev. Neurosci.*, 3, 177, 1985.
9. **Drejer, J., Larsson, O. M., Kvamme, E., Svenneby, G., Hertz, L., and Schousboe, A.,** Ontogenetic development of glutamate metabolizing enzymes in cultured cerebellar granule cells and in cerebellum, *Neurochem. Res.*, 10, 49, 1985.
10. **Kvamme, E., Schousboe, A., Hertz, L., Torgner, I. Aa., and Svenneby, G.,** Developmental change of endogenous glutamate and gamma-glutamyl transferase in cultured cerebral cortical interneurons and cerebellar granule cells and in mouse cerebral cortex and cerebellum *in vivo*, *Neurochem. Res.*, 10, 993, 1985.
11. **Hertz, L., Juurlink, B. H. J., and Szuchet, S.,** Cell cultures, in *Handbook of Neurochemistry*, Vol. 8, 2nd ed., Lajtha, A., Ed., Plenum Press, New York, 1985, 603.
12. **Hertz, L., Juurlink, B. H. J., Szuchet, S., and Walz, W.,** Cell and tissue culture, in *Neuromethods*, Vol. 1, Boulton, A. A. and Baker, G. B., Eds., Humana Press, Clifton, N.J., 1985, 117.
13. **Schousboe, A.,** Differences between astrocytes in primary cultures and glial cell lines in uptake and metabolism of putative amino acid transmitters, in *Cell, Tissue and Organ Cultures in Neurobiology*, Fedoroff, S. and Hertz, L., Eds., Academic Press, New York, 1977, 441.
14. **Hertz, L., Yu, A. C. H., Potter, R. L., Fisher, T. E., and Schousboe, A.,** Metabolic fluxes from glutamate and towards glutamate in neurons and astrocytes in primary cultures, in *Glutamine, Glutamate, and GABA in the Central Nervous System*, Hertz, L., Kvamme, E., McGeer, E. G., and Schousboe, A., Eds., Alan R. Liss, New York, 1983, 327.
15. **Patel, M. S.,** The relative significance of CO_2-fixing enzymes in the metabolism of rat brain, *J. Neurochem.*, 22, 717, 1974.
16. **Roberts, E.,** Strategies for identifying sources and sites of formation of GABA-precursor or transmitter glutamate in brain, in *Glutamate as a Neurotransmitter*, Di Chiara, G. and Gessa, G. L., Eds., Raven Press, New York, 1981, 91.
17. **McGeer, E. G., McGeer, P. L., and Thompson, S.,** GABA and glutamate enzymes, in *Glutamine, Glutamate and GABA in the Central Nervous System*, Hertz, L., Kvamme, E., McGeer, E. G., and Schousboe, A., Eds., Alan R. Liss, New York, 1983, 3.
18. **Thompson, S. G., Wong, P. T.-H., Leong, S. F., and McGeer E. G.,** Regional distribution in rat brain of l-pyrroline-5-carboxylate dehydrogenase and its localization to specific glial cells, *J. Neurochem.*, 45, 1791, 1985.
19. **Meister, A.,** Enzymology of amino acid transport, *Science*, 180, 33, 1974.
20. **Minn, A. and Besagni, D.,** Uptake of L-glutamine into synaptosomes, *Life Sci.*, 33, 225, 1983.
21. **Lisy, V., Stastny, F., Murphy, S., and Hajkova, B.,** Glutamate uptake into cerebral cortex slices is reduced in the presence of a gamma-glutamyl transpeptidase inhibitor, *Experientia*, 39, 111, 1983.
22. **Patel, A. J., Hunt, A., Gordon, R. D., and Balazs, R.,** The activities in different neural cell types of certain enzymes associated with the metabolic compartmentation of glutamate, *Dev. Brain Res.*, 4, 3, 1982.
23. **Drejer, J. and Schousboe, A.,** Ornithine-δ-amino transferase exhibits different kinetic properties in astrocytes, cerebral cortex interneurons, and cerebellar granule cells in primary cultures, *J. Neurochem.*, 42, 1194, 1984.
24. **Yu, A. C. H., Drejer, J., Hertz, L., and Schousboe, A.,** Pyruvate carboxylase activity in primary cultures of astrocytes and neurons, *J. Neurochem.*, 41, 1484, 1983.
25. **Hauser, K. and Bernasconi, R.,** Rat cortical neurons in dissociated cell culture: changes in GABA and guanyl cyclase activity during development, in *Tissue Culture in Neurobiology*, Giacobini, E., Vernadakis, A., and Shahar, A., Eds., Raven Press, New York, 1980, 205.
26. **Borg, J., Spitz, B., Hamel, G., and Mark, J.,** Selective culture of neurons from rat cerebral cortex: morphological characterization, glutamate uptake and related enzymes during maturation in various culture media, *Dev. Brain Res.*, 18, 37, 1985.
27. **Snodgrass, S. R., White, W. F., Biales, B., and Dichter, M.,** Biochemical correlates of GABA function in rat cortical neurons in culture, *Brain Res.*, 190, 123, 1980.
28. **Schousboe, A., Svenneby, G., and Hertz, L.,** Uptake and metabolism of glutamate in astrocytes cultured from dissociated mouse brain hemispheres, *J. Neurochem.*, 29, 999, 1977.

29. **Schousboe, A., Hertz, L., Svenneby, G., and Kvamme, E.,** Phosphate activated glutaminase activity and glutamine uptake in primary cultures of astrocytes, *J. Neurochem.*, 32, 943, 1979.
30. **Wilson, S. H., Schrier, B. K., Farber, J. L., Thompson, E. J., Rosenberg, R. N., Blume, A. J., and Nirenberg, M. W.,** Markers for gene expression in cultured cells from the nervous system, *J. Biol. Chem.*, 247, 3159, 1972.
31. **Shine, H. D., Hertz, L., De Vellis, J., and Haber, B.,** A fluorimetric assay for gamma-glutamyl transpeptidase: demonstration of enzymatic activity in cultured cells of neural origin, *Neurochem. Res.*, 6, 453, 1981.
32. **Roth-Schechter, B. F., Laluet, M., Tholey, G., and Mandel, P.,** The effect of pentobarbital on the carbohydrate metabolism of glial cells in culture, *Biochem. Pharmacol.*, 26, 1307, 1977.
33. **Kvamme, E., Svenneby, G., Hertz, L., and Schousboe, A.,** Properties of phosphate activated glutaminase in astrocytes cultured from mouse brain, *Neurochem. Res.*, 7, 761, 1982.
34. **Tardy, M., Fages, C., Rolland, B., Bardakjian, J., and Gonnard, P.,** Effect of prostaglandins and dibutyryl cyclic AMP on the morphology of cells in primary astroglial cultures and on metabolic enzymes of GABA and glutamate metabolism, *Experientia*, 37, 19, 1981.
35. **Bardakdjian, J., Tardy, M., Pimoule, C., and Gonnard, P.,** GABA metabolism in cultured glial cells, *Neurochem. Res.*, 4, 517, 1979.
36. **Juurlink, B. H. J., Schousboe, A., Jørgensen, O. S., and Hertz, L.,** Induction by hydrocortisone of glutamine synthetase in mouse primary astrocyte cultures, *J. Neurochem.*, 36, 136, 1981.
37. **Hallermayer, K., Harmening, C., and Hamprecht, B.,** Cellular localization and regulation of glutamine synthetase in primary cultures of brain cells from newborn mice. *J. Neurochem.* 37, 43, 1981.
38. **Patel, A. J. and Hunt, A.,** Observations on cell growth and regulation of glutamine synthetase by dexamethazone in primary cultures of forebrain and cerebellar astrocytes, *Brain Res.*, 350, 175, 1985.
39. **Tholey, G., Ledig, M., Bloch, S., and Mandel, P.,** Glutamine synthetase and energy metabolism enzymes in cultured chick glial cells: modulation by dibutyryl cyclic AMP, hydrocortisone, and trypsinization, *Neurochem. Res.*, 10, 191, 1985.
40. **Wenthold, R. J. and Altschuler, R. A.,** Immunocytochemistry of aspartate aminotransferase and glutaminase, in *Glutamine, Glutamate and GABA in the Central Nervous System*, Hertz, L., Kvamme, E., McGeer, E. G., and Schousboe, A., Eds., Alan R. Liss, New York, 1983, 33.
41. **Schousboe, A. and Hertz, L.,** Primary cultures of GABAergic and glutamatergic neurons as a model system to study neurotransmitter function. II. Developmental aspects, in *Model Systems of Development and Aging of the Nervous System*, Vernadakis, A., Privat, A., Lauder, J. N., Tiniras, P. S., and Giacobini, E., Eds., Kluwer-Nijhoff, Boston, 1986, 33.
42. **Oertel, W. H., Schmechel, D. E., Mugnaini, E., Tappaz, M. L., and Kopin, I. J.,** Immunocytochemical localization of glutamate decarboxylase in rat cerebellum with a new antiserum, *Neuroscience*, 6, 2715, 1981.
43. **Kvamme, E., Svenneby, G., Torgner, I. Aa., Drejer, J., and Schousboe, A.,** Postnatal development of glutamate metabolizing enzymes in hippocampus from mice, *Int. J. Dev. Neurosci.*, 3, 359, 1985.
44. **Norenberg, M. D. and Martinez-Hernandez, A.,** Fine structural localization of glutamine synthetase in astrocytes of rat brain, *Brain Res.*, 161, 303, 1979.
45. **Vaccaro, D. E., Leeman, S. E., and Reif-Lehrer, L.,** Glutamine synthetase activity in vivo and in primary cell cultures of rat hypothalamus, *J. Neurochem.*, 33, 953, 1979.
46. **Shank, R. P., Campbell, G. L., Freyteg, S. D., and Utter, M. F.,** Evidence that pyruvate carboxylase is an astrocyte specific enzyme in CNS tissues, *Abstr. Soc. Neurosci.*, 7, 936, 1981.
47. **Shank, R. P., Bennet, G. S., Freyteg, S. D., Campbell, G. L., and Utter, M. F.,** Pyruvate carboxylase: an astrocyte-specific enzyme implicated in the replenishment of amino acid neurotransmitter pools, *Brain Res.*, 329, 364, 1985.
48. **Schousboe, A., Drejer, J., and Hertz, L.,** Comparison of the metabolism of glucose and glutamate in cultured cerebellar granule cells, *J. Neurochem.*, Suppl. 44, 168, 1985.
49. **Yu, A. C. H., Fisher, T. E., Hertz, E., Tildon, J. T., Schousboe, A., and Hertz, L.,** Metabolic fate of [^{14}C]-glutamine in mouse cerebral neurons in primary cultures, *J. Neurosci. Res.*, 11, 351, 1984.
50. **Yu, A. C., Schousboe, A., and Hertz, L.,** Metabolic fate of ^{14}C-labeled glutamate in astrocytes in primary cultures, *J. Neurochem.*, 39, 954, 1982.
51. **Yu, A. C. H., Schousboe, A., and Hertz, L.,** Influence of pathological concentrations of ammonia on metabolic fate of ^{14}C-labeled glutamate in astrocytes in primary cultures, *J. Neurochem.*, 42, 594, 1984.
52. **Yu, A. C. H. and Hertz, L.,** Metabolic sources of energy in astrocytes, in *Glutamine, Glutamate, and GABA in the Central Nervous System*, Hertz, L., Kvamme, E., McGeer, E. G., and Schousboe, A., Eds., Alan R. Liss, New York, 1983, 431.
53. **Lopes-Cardozo, M., Larsson, O. M., and Schousboe, A.,** Acetoacetate and glucose as lipid precursors and energy substrates in primary cultures of astrocytes and neurons from mouse cerebral cortex, *J. Neurochem.*, 46, 773, 1986.

54. **Hertz, L. and Schousboe, A.,** Role of astrocytes in compartmentation of amino acid and energy metabolism, in *Astrocytes* Vol. 2, Fedoroff, S. and Vernadakis, A., Eds., Academic Press, New York, 1986, 179.
55. **Lai, J. C. K., Murthy, Ch. R. K., Hertz, L., and Cooper, A. J. L.,** NH_3 and β-methyleneaspartate inhibit neuronal and glial glutamate oxidation, *Trans. Am. Soc. Neurochem.*, 17, 217, 1986.
56. **Martin, D. L., Waniewski, R. A., and Miller, M. S.,** Conversion of glutamate to glutamine by rat cortical astrocytes, *Abstr. 14th Annu. Meet. Neurosci.*, Anaheim, 765, 1984.
57. **Hertz, L., Murthy, Ch. R. K., Lai, J. C. K., Shargool, M., Fitzpatrick, S. M., Yu, A. C. H., Schousboe, A., and Cooper, A. J. L.,** Metabolic effects of ammonia on astrocytes in primary cultures, *Trans. Am. Soc. Neurochem.*, 17, 335, 1986.
58. **Tapia, R.,** Biochemical pharmacology of GABA in CNS, in *Handbook of Psychopharmacology*, Vol. 4, Iversen, L. L., Iversen, S. D. and Snyder, S. H., Eds., Plenum Press, New York, 1975, 1.
59. **Waniewski, R. A. and Martin, D. L.,** Exogenous glutamate is metabolized to glutamine and exported by rat primary astrocyte cultures, *J. Neurochem.*, 47, 304, 1986.
60. **Yu, A. C. H., Hertz, E., Schousboe, A., and Hertz, L.,** Uptake and metabolism of malate in cultures of astrocytes and neurons, *Trans. Am. Soc. Neurochem.*, 15, 216, 1984.
61. **Ramaharobandro, N., Borg, J., Mandel, P., and Mark, J.,** Glutamine and glutamate transport in cultured neuronal and glial cells, *Brain Res.*, 244, 113, 1982.
62. **Potter, R. L., Yu, A. C., Schousboe, A., and Hertz, L.,** Metabolic fate of [U-^{14}C] labeled glutamate in primary cultures of mouse astrocytes as a function of development, *Dev. Neurosci.*, 5, 278, 1982.
63. **Borg, J., Hamel, G., Spitz, B., and Mark, J.,** Compartmentation of glutamate and glutamine metabolism in cultured nerve cells, in *Glutamine, Glutamate and GABA in the Central Nervous System*, Hertz, L., Kvamme, E., McGeer, E. G. and Schousboe, A., Eds., Alan R. Liss, New York, 1983, 317.
64. **Yudkoff, M., Nissim, I., Pleasure, D., Kim, S., Hummeler, K., and Segal, S.,** Ammonia and amino acid interaction in cultured brain cells: studies with $^{15}NH_3$, in *Glutamine, Glutamate and GABA in the Central Nervous System*, Hertz, L., Kvamme, E., McGeer, E. G., and Schousboe, A., Eds., Alan R. Liss, New York, 1983, 389.
65. **Yudkoff, M., Nissim, I., Kim, S. U., Pleasure, D., and Segal, S.,** Metabolism of $^{15}NH_3$ in organotypic cerebellar explants and cultured astrocytes: studies with gas chromatography-mass spectrometry, *J. Neurochem.*, 43, 283, 1984.
66. **Cotman, C. W., Foster, A., and Lanthorn, T.,** An overview of glutamate as a neurotransmitter, in *Glutamate as a Neurotransmitter*, Di Chiara, G. and Gessa, G. L., Eds., Raven Press, New York, 1981, 1.
67. **Bradford, H. F., Ward, H. K., and Thanki, C. M.,** Glutamine as a neurotransmitter precursor: complementary studies in vivo and in vitro on the synthesis and release of transmitter glutamate and GABA, in *Glutamine, Glutamate, and GABA in the Central Nervous System*, Hertz, L., Kvamme, E., McGeer, E. G., and Schousboe, A., Eds., Alan R. Liss, New York, 1983, 249.
68. **Yu, A. C. H., Schousboe, A., and Hertz, L.** unpublished experiments.
69. **Yu, A. C. H.,** Interaction Between Neurons and Astrocytes in Glutamate and Glutamine Metabolism, Ph.D. thesis, University of Saskatchewan, Saskatoon, 1984.
70. **Shank, R. P. and Campbell, G. L.,** α-Ketoglutarate and malate uptake and metabolism by synaptosomes: further evidence for an astrocyte to neuron metabolic shuttle, *J. Neurochem.*, 42, 1153, 1984.
71. **Shank, R. P. and Campbell, G. L.,** Glutamine, glutamate and other possible regulators of α-ketoglutarate and malate uptake by synaptic terminals, *J. Neurochem.*, 42, 1162, 1984.
72. **LaNoue, K. F. and Schoolwerth, A. C.,** Metabolite transport in mitochondria, *Annu. Rev. Biochem.*, 48, 871, 1979.
73. **Wong, P. T.-H., McGeer, E. G., and McGeer, P. L.,** A sensitive radiometric assay for ornithine aminotransferase: regional and subcellular distributions in rat brain, *J. Neurochem.*, 36, 501, 1981.
74. **Yoneda, Y. and Roberts, E.,** Synaptosomal biosynthesis of GABA from ornithine and its feedback inhibition by GABA, in *Problems in GABA Research from Brain to Bacteria*, Excerpta Medica, Amsterdam, 1982, 55.
75. **Yoneda, Y., Roberts, Y., and Dietz, G. W., Jr.,** A new synaptosomal biosynthetic pathway of glutamate and GABA from ornithine and its negative feedback inhibition by GABA, *J. Neurochem.*, 38, 1688, 1982.
76. **Shank, R. P. and Campbell, G. L.,** Amino acid uptake, content and metabolism by neuronal and glial enriched cellular components from mouse cerebellum, *J. Neurosci.*, 4, 58, 1984.
77. **Bradford, H. F., Ward, K. H., and Thomas, A. J.,** Glutamine as a substrate for nerve endings, *J. Neurochem.*, 30, 1453, 1978.
78. **Hamberger, A. C., Chiang, G. H., Nylen, E. S., Scheff, S. W., and Cotman, C. W.,** Glutamate as a CNS transmitter. I. Evaluation of glucose and glutamine as precursors for the synthesis of preferentially released glutamate, *Brain Res.*, 168, 513, 1979.

79. **Hamberger, A., Berthold, C.-H., Karlsson, B., Lehman, A., and Nystrom, B.,** Extracellular GABA, glutamate and glutamine *in vivo* — perfusion dialysis of the rabbit hippocampus, in *Glutamate, Glutamine, and GABA in the Central Nervous System,* Hertz, L., Kvamme, E., McGeer, E. G., and Schousboe, A., Eds., Alan R. Liss, New York, 1983, 473.
80. **Duffy, T. E., Plum, F., and Cooper, A. J. L.,** Cerebral ammonia metabolism in vivo, in *Glutamine, Glutamate and GABA in the Central Nervous System,* Hertz, L., Kvamme, E., McGeer, E. G., and Schousboe, A., Eds., Alan, R. Liss, New York, 1983, 371.
81. **Murthy, Ch. R. K. and Hertz, L.,** unpublished experiments.
82. **Murthy, Ch. R. K., Hertz, E., and Hertz, L.,** Ammonia effects on oxidation of branched chain amino acids by astrocytes, *Trans. Am. Soc. Neurochem.,* 17, 125, 1986.
83. **Jessy, J. and Murthy, Ch. R. K.,** Elevation of transamination of branched chain amino acids in brain in acute ammonia toxicity, *Neurochem. Int.,* 7, 1027, 1985.
84. **Tildon, J. T. and Roeder, L. M.,** Glutamine oxidation by dissociated cells and homogenates of rat brain: kinetics and inhibitor studies, *J. Neurochem.,* 42, 1060, 1984.
85. **Zielke, H. R., Zielke, C. L., and Ozand, P. T.,** Glutamine, a major energy source for cultured mammalian cells, *Fed. Proc., Fed. Am. Soc. Exp. Biol.,* 43, 121, 1984.
86. **Windmueller, H. G. and Spaeth, A. E.,** Uptake and metabolism of plasma glutamine by the small intestine, *J. Biol. Chem.,* 249, 5070, 1974.
87. **Drejer, J., Schousboe, A., and Hertz, L.,** unpublished experiments.
88. **Yu, A., Schousboe, A., and Hertz, L.,** unpublished.
89. **Drejer, J., Schousboe, A., Shargool, M., and Hertz, L.,** unpublished.

Chapter 18

TRANSMITTER GLUTAMATE IN MAMMALIAN HIPPOCAMPUS AND STRIATUM

Frode Fonnum

TABLE OF CONTENTS

I. INTRODUCTION

The acidic amino acids, aspartate and glutamate, are probably quantitatively the most important excitatory transmitters in the central nervous system.[1,2] The evidence for glutamate as a neurotransmitter in brain has been discussed in several reviews.[1] (See also Chapter 15.) Much of the evidence for its transmitter role has been provided through studies in two brain regions, namely, the hippocampus and the neostriatum. The object of this review is therefore to focus on the work on glutamate transmitter function from these two brain regions.

The main difficulty in studying the role of glutamate as a transmitter or, even more, to identify the transmitter pool of glutamate is the widespread role of glutamate in brain metabolism. The methods available to identify glutamatergic pathways in the brain have been extensively discussed.[3] The following properties of a glutamatergic terminal are most commonly used for its identification:

1. The high-affinity uptake of L-glutamate and D-aspartate assayed either biochemically in an homogenate or autoradiographically in a slice, including the changes in high-affinity uptake following lesions of pathways
2. The decrease in endogenous glutamate as a result of degeneration of glutamatergic terminals
3. The Ca^{2+}-dependent release of either endogenous glutamate, exogenously added L-glutamate or D-aspartate, or newly synthesized L-glutamate from glutamine or glucose from synaptosomes or slices after chemical depolarization or electrical stimulation
4. The blockade of specific postsynaptic responses by glutamate antagonists

In this paper, the application of these methods to identify the glutamatergic pathways to these two brain regions will be extensively discussed. In addition, several examples will be given on how hippocampus and neostriatum have been used as models to throw further light on different aspects of the transmitter function of glutamate.

II. HIPPOCAMPUS

The hippocampal formation is phylogenetically an old part of the cerebral cortex. It is a laminar structure where the different fiber connections are organized in a well-defined pattern. The hippocampus proper consists of a cellular layer of pyramidal cells and a cellular layer of granular cells in the area dentata (Figure 1).

The area dentata is the major entrance of fibers to the hippocampus. The perforant path, which originates in the enthorinal cortex, terminates in the outer molecular layer of the area dentata where it makes contact with the dendrites of the granular cells.[4,5] The granular cells project their axons, the mossy fibers, to the pyramidal cells in hilus and in the CA3.[6] The CA3 pyramidal cells project the so-called Schaffer collaterals with characteristic enpassage synapses to the CA1 pyramidal cell apical dendrites.[7] The CA3 cells send a second branch of the axon through the fornix/fimbria mainly to the lateral septum.[8] The CA1 pyramidal cells project further to the subiculum, but also send fibers through the fornix/fimbria. The pyramidal cells of the subiculum project to several brain regions such as the basal hypothalamus, nucleus accumbens, and mammillary bodies.[8]

Another characteristic feature of the hippocampus is the lamellar organization of the structure. Thus, the perforant path, the mossy fibers, and the Schaffer collaterals are organized in thin lamellae.[9] This has allowed the development of the hippocampal slice preparation[10] which has been very valuable in revealing the electrophysiological events in the hippocampus and in studying the release of glutamate after specific stimulation.[11-13]

Kainic acid treatment destroyed the pyramidal cells and the degeneration was accompanied

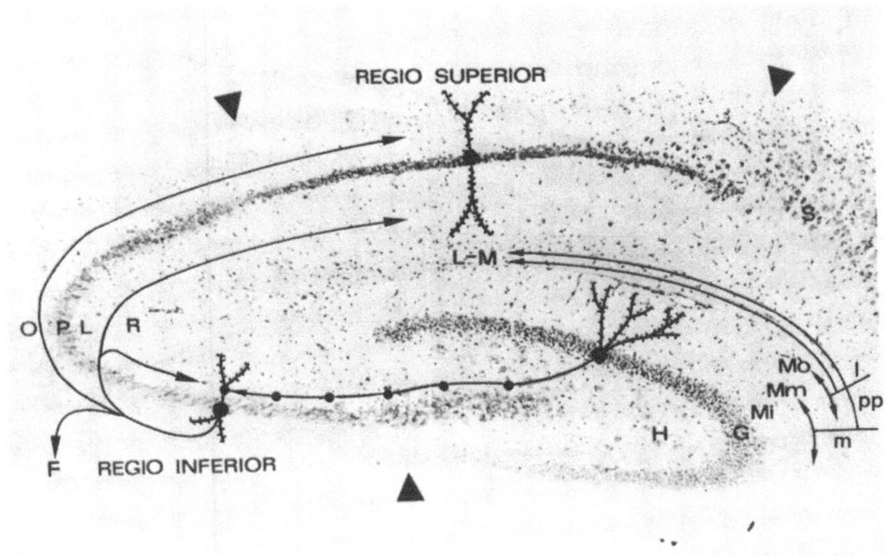

FIGURE 1. A section of normal hippocampus stained with toluidine blue. G: granular cells; H: hilus fasciae dentatae; Mo, Mm, and Mi: outer, middle, and inner part of the molecular layer of area dentata; O: stratum oriens; P: stratum pyramidale; R: stratum radiatum; S: subiculum; L-M: laccinosum moleculare of hippocampus; and L: stratum lucidum (mossy fiber layer). In the drawing is shown the three-link chain of excitatory neurons: perforant path axons (pp) with the lateral and medial parts (l, m), the granular cell axons (the mossy fiber), and the pyramidal cell axons from CA3 (regio inferior) to fimbria and to stratum radiatum and stratum ariens (regio superior).

by a large decrease in D-aspartate uptake and in the endogenous glutamate level in the hippocampus.[14] This indicated that a large part of the neurons could be glutamatergic.

A. Perforant Path

The perforant path projects to the outer molecular layer of the area dentata and the molecular layer of hippocampus. These layers have by autoradiography been shown to have a high-affinity uptake of L-glutamate, L-aspartate, and D-aspartate.[15-17] The uptake is less dense than for the neighboring layers (Figure 2). Electron microscopic studies have shown, however, that most of the labeled material has been taken up into axons and nerve terminals.[18] In agreement with a specific localization of the glutamate uptake to the perforant path terminals, the uptake was reduced after transection of the perforant path[15,16] when examined by both autoradiographical and biochemical methods.

Endogenous glutamate and aspartate were released from dentatae slices in a Ca^{2+}-dependent manner by chemical or electrical depolarization, and the release was decreased after cutting the perforant path.[19,20] Release of newly synthesized glutamate from both glucose and glutamine showed a similar behavior.[21] There are reports showing a decrease in both the endogenous aspartate[22] or the synaptosomal content of glutamate[23] after destruction of the perforant path.

Pharmacological studies showed a preferential block both to the stimulation of perforant path by glutamate antagonists such as glutamate diethylester, piperazine-2,3-dicarboxylic acid, and 2-amino-5-phosphonobutyrate.[24-28] It is, however, interesting to note that the pharmacological response of 2-amino-5-phosphonobutyrate, 2-amino-4-phosphonobutyrate, and baclofen was different in the lateral and medial perforant path.[29,30] The perforant path satisfies all the four criteria listed in the introduction for having glutamate as its neurotransmitter.

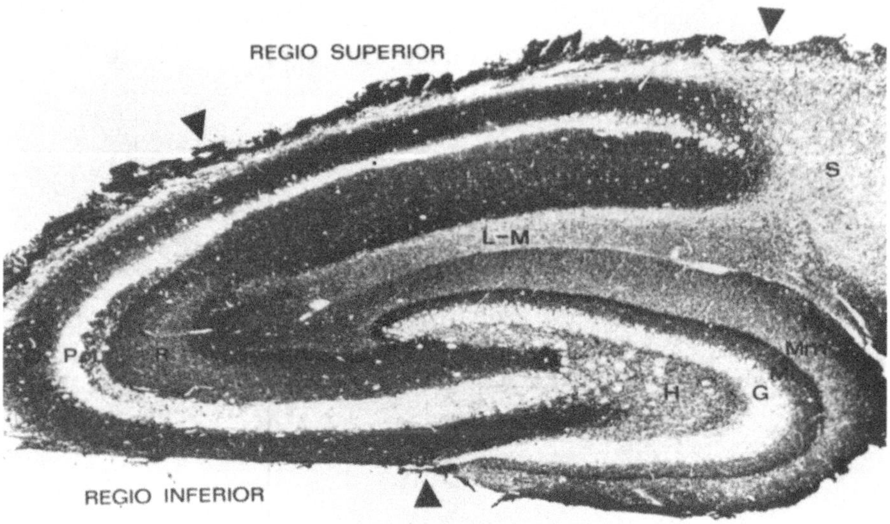

FIGURE 2. Autoradiography of high-affinity uptake of [^{3}H]glutamate in rat hippocampus. The uptake is distributed in a laminar pattern and the highest silver grain density is seen in strata oriens and radiatum of hippocampus and the inner molecular layer of area dentata.

B. Mossy Fibers

The mossy fibers show a specific labeling with D-aspartate or L-glutamate at the electron microscopic level.[18] The uptake was not very dense compared with the other layers, but it became very distinct after kainic acid treatment, which removes the pyramidal cells in CA3 and thereby the labeling in surrounding structures.[17] Also, retrograde transport of D-aspartate has been shown to occur from the mossy fiber region to the granular cells.[31] Since the mossy fibers occupy such a small part of the CA3 region, it has not been possible to demonstrate any effect on release and uptake of glutamate and aspartate in CA3 region after destruction of the mossy fibers.[20] The reason for the scattered uptake of D-aspartate in the mossy fibers may be due to the mechanical disruption of these large terminals in a slice or that only a part of them are glutamatergic/aspartatergic. The mossy fibers have been demonstrated to also contain enkephalin.[32,33,71] The mossy fibers also show heavy TIM staining (TIMs sulfide silver method) and have a high concentration of Zn.[34] Stimulation of granule cells through the activation of the perforant path was accompanied by a reduction in TIM staining of the mossy fiber.[35] This has given rise to speculation on a coupling between Zn and transmitter glutamate, particularly during epilepsy.

C. CA3 Axon Collaterals

Autoradiographic pictures after high-affinity uptake of D-aspartate showed a high density of labeling both in the stratum radiatum of CA1 where the Schaffer collaterals terminate (Figure 2) and in the lateral septum where the CA3 axons terminate.[36] In agreement, surgical lesion of the CA3/4 pyramidal cells leads to a substantial decrease in high-affinity glutamate uptake in the stratum radiatum of CA1.[15]

Using the transverse hippocampal slice, devoid of the area dentata, it was possible to show a Ca^{2+}-dependent release of exogenously added [^{3}H]-D-aspartate after stimulation of the Schaffer collaterals, whereas L-leucine and GABA were not specifically released.[11,37] Electrical stimulation of the Schaffer collaterals was accompanied by the release both of endogenous aspartate and glutamate.[38] Electrical stimulation of the Schaffer collaterals has been used to show that high-affinity uptake increased during stimulation.[39] This observation is of great interest with regard to the glutamatergic mechanism.

The collaterals of the Schaffer fibers project to the lateral, but not the medial, septum. In agreement, autoradiography of lateral, but not medial, septum slices showed dense labeling after high-affinity uptake of D-aspartate.[36] Bilateral transection of the fibers from CA3 leads to a decrease in the high-affinity uptake of D-aspartate in the lateral, but not medial, septum.[14,40,41] The lesion was also accompanied by a loss of endogenous glutamate, but not aspartate or any other amino acid.[14] Exogenously added D-aspartate was also released in a Ca^{2+}-dependent manner from the septal slice upon electrical stimulation of the fimbrial.[12] The response of glutamate to pyramidal cells in CA1 and the response of glutamate antagonists to stimulation of the Schaffer collateral also show that this pathway is glutamatergic.[42-44]

D. CA1 Fibers

The major projection from CA1 pyramidal cells is directed toward the subiculum. This region has less glutamate uptake than the other regions of hippocampus (Figure 2), but it is decreased after transection of the CA1 fibers.[15] These fibers are therefore probably also glutamatergic/aspartatergic.

E. Fornix/Fimbria Fibers

Most fibers in the descending fornix originate from the pyramidal cells of the subiculum.[8] They have a widespread distribution and terminate mainly in the bed nucleus of stria terminals, anterior diagonal band nucleus, nucleus accumbens, basal hypothalamus, and mammillary body. Transection of fimbria/fornix was accompanied by a decrease in both high-affinity uptake of D-aspartate and in the endogenous level of glutamate, but not of the other amino acids except aspartate to a small extent.[45] Glutamate is therefore a strong candidate as neurotransmitter for these fibers. It is interesting to note that stimulation of the hippocampus increases the deoxyglucose utilization and therefore the metabolic activity in these regions.

The results therefore indicate that glutamate is possibly the main activator of the cholinergic cells in the anterior nucleus of the diagonal band.[45,46] It was found that the other cholinergic cell groups do not receive a similar large glutaminergic input.[46a]

The fornix fibers in the basal hypothalamus project to the arcuate and ventromedial nucleus.[47] These cells are sensitive to glutamate and they release newly synthesized glutamate on stimulation.[48] These cells, particularly the cholinergic cells, are very sensitive to the neonatal toxic effect of glutamate.[49] It is also interesting that non-toxic doses of glutamate lead to rapid changes in the release pattern of the pituitary gland.[50] Transmitter glutamate may therefore play an important role in the regulation of the endocrine function in hypothalamus.

F. Strategy for Studying Glutamatergic Mechanism in Hippocampus

The distinct pattern for high-affinity glutamate uptake in the hippocampus can be taken as a distinct pattern for the distribution of glutamate terminals. It is therefore of great interest to compare the histochemical pattern of any other biochemical component with that of high-affinity glutamate uptake. Immunocytochemical mapping of glutamate-like activity has been compared with that of high-affinity glutamate. After special treatment of the slices, a striking similarity occurs. This has been taken as evidence that the glutamate-like immunohistochemistry could be a valuable tool for identifying glutamatergic structures.[51] Similar comparisons have been carried out with immunohistochemistry of the soluble aspartate aminotransferase or glutaminase.[52] Advantage of this pattern has also been taken for comparison with the total aspartate aminotransferase activity and glutamate dehydrogenase.[53-55]

Long-term stimulation is a phenomenon in the hippocampus which has been linked to

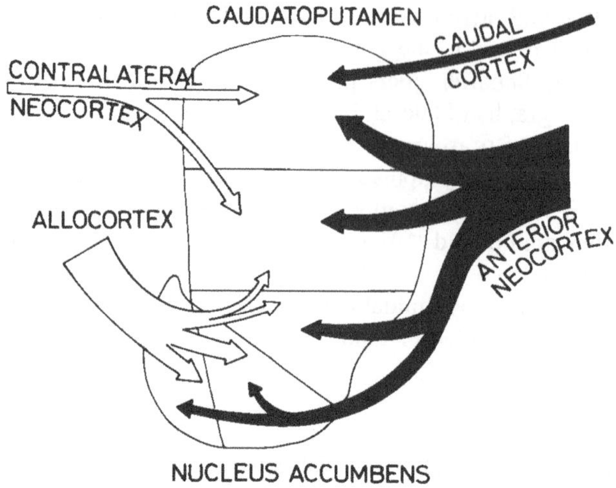

FIGURE 3. Glutamatergic inputs to the rostral part of caudato putamen and nucleus accumbens.[66] The major glutamatergic input to caudatoputamen comes from the ipsilateral anterior neocortex with minor inputs from the caudal neocortex, contralateral neocortex, and from allocortex. The major input to nucleus accumbens is mainly derived from allocortex and the anterior neocortex.

memory function. The activation of long-term potentiation is similar to the increase in glutamate release[11] and probably linked to the *N*-methyl-D-aspartate (NMDA) receptors.[56] This was first observed with the Schaffer collaterals where long-term potentiation occurs, but not with the septal-fimbria fibers where it does not occur. This finding has been further substantiated and expanded to comprise the long-term potentiation observed in the perforant path.[56]

Studies of the glutamatergic structures in the hippocampus have been important for revealing the mechanism of kainic acid toxicity. In this case, the effect of interruption of well-defined pathways has been important.[57] Hippocampus slices have been particularly useful in investigating glucose and glutamine as precursors for glutamate transmitter synthesis.[28,58]

III. NEOSTRIATUM (CAUDATE NUCLEUS AND PUTAMEN)

Both biochemical and autoradiographical techniques showed a high activity of high-affinity uptake of D-aspartate in neostriatum.[36,59,60] There is a distinct somatotopical arrangement of cortical fibers projecting to the neostriatum.[61-64] The loss of glutamate uptake was well correlated to the disruption of these cortical fibers (Figure 3).[59,65,66] A small part of the glutamate input to the most ventral part of the neostriatum may be derived from hippocampal formation. Cortical ablation was also accompanied by a 20 to 30% decrease in endogenous glutamate and a small decrease in the aspartate level, which was not always significant, and an increase in the glutamine level.[59,67-69]

Injection of labeled D-aspartate in neostriatum showed that the radiolabel was transported to the neocortex.[70] This technique is based on a specific uptake of D-aspartate in vivo into glutamatergic terminals and its retrograde transport to soma in the neocortex.

The release of exogenous glutamate, endogenous glutamate, or glutamate synthesized from exogenous glutamine or glucose from neostriatal slices during potassium depolarization was reduced after decortication.[71-75] The excitatory response to cortical stimulation could be suppressed by the iontophoretic application of the glutamate antagonist glutamate diethylester.[76] The corticostriatal pathway therefore satisfies all the four criteria for glutamate as a neurotransmitter. This region has become important for studying different aspects of transmitter glutamate.

A. Strategy for Studying Glutamatergic Mechanism in Neostriatum

The glutamatergic input to neostriatum is well documented.[72] Unilateral decortication will give, in the same animal, an intact neostriatum and a neostriatum deprived of its glutamatergic input. This is an almost ideal situation for studying the composition and function of glutamatergic terminals. Extensive use has been made of the neostriatal model both for quantification of the different glutamate pools, for studying the turnover of these pools, and for studying the composition of the glutamatergic terminals.[2,72,77]

B. Glutamate Pools in Neostriatum

In neostriatum, as in all other regions of the brain, there are at least four pools of glutamate. They are the transmitter pool, the precursor pool for GABA synthesis, the pool of glutamate in glial cells, and the metabolic pool of glutamate in the nonglutamatergic neurons.[2]

The transmitter pool of glutamate can be determined after selective degeneration of glutamatergic terminals. In the neostriatum, this pool therefore represents the decrease in the level of endogenous glutamate after decortication. This value ranges from 20 to 50%, but most figures are between 20 to 30%.[59,67-69,77,78] Electron microscopic studies of neostriatum after cortical ablation have shown that degenerating terminals only occupy a few percent of the total neostriatal area. (Fonnum et al. to be published). The concentration of glutamate in glutamatergic terminals is therefore probably very high.

The pool of glutamate which functions as a precursor for GABA can be estimated in a similar way after selective destruction of GABAergic terminals. In substantia nigra, the region with the highest content of GABAergic terminals, selective destruction of GABAergic terminals was accompanied by a small decrease in glutamate level (between 0 to 1 μmol/g).[79,80] The precursor pool of glutamate for GABA synthesis if therefore probably less than 5% of the total glutamate pool (10 μmol/g) in neostriatum, where the GABA level is less than half that of substantia nigra. In agreement, the level of glutamate is much higher in neostriatum than in globus pallidus, which is a GABA-rich and transmitter-glutamate poor region. (Paulsen et al., to be published). The concentration of glutamate in GABAergic terminals is therefore relatively low.

The proportion of glutamate in the glial cells can be estimated from the relative specific radioactivity of glutamine compared to glutamate after administering precursors which specifically label the glial cells. In several regions of the cat brain, including neostriatum, this ratio with labeled acetate was 5, which should indicate that the glial pool is 20% of the total glutamate pool.[81] This is probably an overestimation since acetate also enters the neuronal pool.

Intrastriatal injection of fluorocitrate reduced the glutamine content 80% and glutamate content 30 to 40% after 4 hr. This lesion specifically affects the glial cells.[82] The reduction in glutamate level could be an overestimation and comprises both the reduction in glial glutamate and also a reduction in terminal glutamate secondary to the loss of its precursor glutamine.

The main pool glutamate in the brain is probably linked to the general metabolism in neurons. In neostriatum, destruction of neuronal cell bodies with kainic acid was accompanied by 50% loss of glutamate.[83] This is expected to account for the glutamate content in neuronal structures except glutamatergic and monoaminergic terminals.

In summary, in neostriatum the glutamate transmitter pool is 20 to 30% the precursor pool for GABA is about 5%, the glial pool 10 to 30%, and the metabolic pool for glutamate about 50%.

C. Turnover of Glutamate in Neostriatum

The neostriatal model has also been used to study the biochemical changes in the intact and decorticated neostriatum accompanying hypoglycemia or treatment with methionine sulfoximine. A general finding is that the turnover of glutamate is higher in the intact than in the decorticated neostriatum. This is particularly apparent during hypoglycemia where the conversion of glutamate, glutamine, and aspartate occurred much faster on the intact side.[67,78] An additional observation which followed was that the toxic effect of hypoglycemia on neurons in neostriatum is reduced by decortication or by the addition of glutamate antagonists.[84,85]

In agreement, infusion of glutamate increased the turnover of GABA in neostriatum.[86] The high turnover of glutamate in the presence of the glutamatergic input may either be caused by transmitter glutamate having a higher turnover than the other glutamate pools or that glutamate fibers are the driving force for neostriatum.

Intrastriatal injections of methionine sulfoximine, an inhibitor of glutamine synthetase, showed that glutamine was reduced faster on the decorticated than the normal side.[77] This was taken as evidence that released glutamate is the most important source for glutamine. Both during hypoglycemia and during methionine sulfoximine treatment, it was observed that a low level of glutamine was able to maintain the glutamate level. The K_m of glutamine for glutamate synthesis is therefore probably low.[67]

D. Composition of the Glutamate Terminal

The difference between decorticated and unoperated neostriatum has been used to study the specific localization of chemical compounds, receptors, and enzymes to glutamatergic terminals in neostriatum.

The specific localization of glutamate and to a lesser extent to aspartate in glutamatergic terminals has already been dealt with. The increase in glutamine level after decortication is thought to reflect an increase in glial structures replacing the degenerated terminals.[68,69]

There was a correlation between high-affinity uptake of [^{3}H] glutamate and high affinity uptake of [^{14}C] 2-oxoglutarate in different brain regions. There was also a significant (40%) decrease in 2-oxoglutarate uptake in neostriatum after decortication.[87] In the same way, there was a correlation between the regional uptake of pyroglutamate and glutamate and a decrease in both in neostriatum after decortication.[88] This shows that glutamate terminals can take up both 2-oxoglutarate and pyroglutamate.

Phosphate-activated glutaminase and ornithine aminotransferase activities were slightly decreased after decortication, but this was not the case with total aspartate aminotransferase or glutamate dehydrogenase activities.[69,89] In contrast, a kainic acid lesion in neostriatum was reported to give large reductions in the activities of glutamate dehydrogenase, aspartate aminotransferase, and GABA aminotransferase as well as in the level of all amino acids.[83,89] The results indicate that there may be a high concentration of glutaminase and ornithine aminotransferase in glutamatergic terminals, whereas glutamate dehydrogenase, aspartate aminotransferase, and GABA aminotransferase are localized to nonglutamatergic structures.

Dopamine receptors have been shown by lesion studies to be localized to the corticostriatal fibers.[90] Further studies showed that it was the number of D-4 dopamine receptors, but not the D-2 receptors, which were decreased on decortication.[91] In agreement, dopamine inhibited the release of glutamate from striatal slices.[75,92] In contrast, glutamate and glutamate agonists increased the dopamine release from striatal slices.[93-96] In this connection it is interesting to note that acute administration of neuroleptics had no effect on the potassium-induced release of glutamate from striatum, but 1-year continuous administration of neuroleptics decreased the release of glutamate significantly.[97]

Met-enkephalin depressed the glutamate-induced responses in striatum. Since it also decreased glutamate binding to membranes, this was believed to be a postsynaptic effect.[98]

The decorticated neostriatum has also been an important model for exploring the mechanism of kainic acid toxicity. The dependency of kainic acid toxicity on an excitatory input was first demonstrated in the decorticated neostriatum.[99] Later it was shown that infusion of glutamate was only toxic in the decorticated neostriatum where the protective uptake mechanisms had been removed.[100] Infusion of glutamate for 1 week continuously caused a significant decrease in the GABAergic and cholinergic parameters.[100]

Recent studies have indicated that glutamate stimulates the formation of inositol phosphate in neostriatum.[101] Further studies may throw light on the link between glutamate and the second messenger.

IV. GENERAL CONCLUSIONS

There is good evidence that several different terminals in the hippocampus used glutamate as their chemical transmitter. The evidence is particularly strong for the perforant path, the Schaffer collateral and fornix/fimbria fibers from subiculum, or the pyramidal cells in CA3, but the evidence also favors the concept that the mossy fibers, the CA1-subiculum connection, and the commissural fibers use glutamate.

The cortical input to neostriatum uses glutamate as a transmitter. This has allowed detailed studies on the glutamate pools, the composition of the glutamate fibers, and the turnover of transmitter glutamate.

REFERENCES

1. **Fonnum, F.,** Glutamate: a neurotransmitter in mammalian brain, *J. Neurochem.*, 42, 1, 1984.
2. **Fonnum, F.** Determination of transmitter amino acid turnover, in *Amino Acids, (Neuromethods,* Vol. 3,) Boulton, A. A., Baker, G. B., and Wood, J. D., Eds., Humana Press, Clifton, N.J., 1985, 201.
3. **Fonnum, F. and Malthe-Sørenssen, D.,** Localization of glutamate neurons, in *Glutamate, Transmitter in the Central Nervous System,* Roberts, P. J., Storm-Mathisen, J., and Johnston, G. A. R., Eds., John Wiley & Sons, New York, 1981, 205.
4. **Andersen, P., Holmquist, B., and Voorhoeve, P. E.,** Entorhinal activation of dentate granule cells, *Acta Physiol. Scand.* 66, 448, 1966.
5. **Andersen, P., Holmquist, B., and Voorhoeve, P. E.,** Excitatory synapses on hippocampal apical dendrites activated by entorhinal activation, *Acta Physiol. Scand.*, 66, 461, 1966.
6. **Blackstad, T. W., Brink, K., Hen, J., and Jeune, B.,** Distribution of hippocampal mossy fibers in the rat. An experimental study with silver staining methods, *J. Comp. Neurol.* 138, 433, 1970.
7. **Andersen, P., Blackstad, T. W., and Lømo, T.,** Location and identification of excitatory synapses on hippocampal pyramidal cells, *Exp. Brain Res.* 1, 236, 1966.
8. **Swanson, L. W. and Cowan, W. M.,** An autoradiographic study of the organization of the efferent connections of the hippocampal formation in the rat, *J. Comp. Neurol.*, 172, 49, 1977.
9. **Andersen, P., Bliss, T. V. P., and Skrede, K. K.,** Lamellar organization of hippocampal excitatory pathways, *Exp. Brain Res.*, 13, 222, 1971.
10. **Skrede, K. K. and Westgaard, R.,** The transverse hippocampal slice: a well-defined cortical structure maintained *in vitro, Brain Res.*, 35, 589, 1971.
11. **Malthe-Sørenssen, D., Skrede, K. K., and Fonnum, F.,** Calcium-dependent release of D-(^{3}H)aspartate evoked by selective electrical stimulation of excitatory afferent fibers to hippocampal pyramidal cells *in vitro, Neuroscience,* 4, 1255, 1979.
12. **Malthe-Sørenssen, D., Skrede, K. K., and Fonnum, F.,** Release of D-(^{2}H)aspartate from the dorsolateral septum after electrical stimulation of the fimbria *in vitro, Neuroscience,* 5, 127, 1980.
13. **Skrede, K. K. and Malthe-Sørenssen, D.,** Increased resting and evoked release of transmitter following repetitive electrical tetanization in hippocampus: a biochemical correlate to long-lasting synaptic potentiation, *Brain Res.*, 208, 436, 1981.
14. **Fonnum, F. and Walaas, I.,** The effect of intrahippocampal kainic acid injections and surgical lesions on neurotransmitters in hippocampus and septum, *J. Neurochem.*, 31, 1173, 1978.

15. **Storm-Mathisen, J.**, Glutamic acid and excitatiory nerve endings: reduction of glutamic acid uptake after axotomy, *Brain Res.*, 120, 379, 1977.
16. **Taxt, T. and Storm-Mathisen, J.**, Uptake of D-aspartate and L-glutamate in excitatory axon terminals in hippocampus: autoradiographic and biochemical comparison with gamma-aminobutyrate and other amino acids in normal rats and in rats with lesions, *Neuroscience*, 11, 79, 1984.
17. **Aamodt, A., Aamböe, A., Walaas, I., Søreide, A. J., and Fonnum, F.**, Autoradiographic demonstration of glutamate structures after stereotaxic injection of kainic acid in rat hippocampus, *Brain Res.*, 294, 341, 1984.
18. **Storm-Mathisen, J. and Iversen, L. L.**, Uptake of D-(^{3}H)glutamic acid in excitatory nerve endings: light and electron microscopic observations in the hippocampal formation of the rat, *Neuroscience*, 4, 1237, 1979.
19. **Nadler, J. V., Vaca, K. W., White, W. F., Lynch, G. S., and Cotman, C. W.**, Aspartate and glutamate as possible transmitters of excitatory hippocampal afferents, *Nature*, 260, 538, 1976.
20. **Nadler, J. V., White, V. F., Vaca, K. W., Perry, B., and Cotman, C. W.**, Biochemical correlates of transmission mediated by glutamate and aspartate, *J. Neurochem.*, 31, 147, 1978.
21. **Hamberger, A., Chiang, G. H., Nylén, E. S., Scheff, S. W., and Cotman, C. W.**, Glutamate as a CNS transmitter. I. Evaluation of glucose and glutamine as precursors for the synthesis of preferentially released glutamate, *Brain Res.*, 168, 513, 1979.
22. **DiLauro, A., Schmid, R. W., and Meek, J. L.**, Is aspartic acid the neurotransmitter of the perforant pathway?, *Brain Res.*, 207, 476, 1981.
23. **Nadler, J. V. and Smith, E. M.**, Perforant path lesion depletes glutamate content of fascia dentata synaptosomes, *Neurosci. Lett.*, 25, 275, 1981.
24. **Freund, R. K., Crooks, S. L., Koerner, J. F., and Johnson, R. L.**, Antagonist activity of phosphorus-containing glutamate analogues in the perforant path, *Brain Res.*, 291, 150, 1984.
25. **Ganary, A. H., Jones, A. W., Watkins, S. C., and Cotman, C. W.**, Parallel antagonism of synaptic transmission and kainate/quisqualate responses in the hippocampus by piperazine-23-dicarboxylic acid analogs, *J. Neurosci.*, 6, 930, 1986.
26. **Hicks, T. P. and McLennan, H.**, Amino acids and the synaptic pharmacology of granule cells in the dentate gyrus of the rat, *Can. J. Physiol. Pharmacol.*, 57, 973, 1979.
27. **Segal, M.**, Glutamate antagonists in rat hippocampus, *Br. J. Pharmacol.*, 58, 341, 1979.
28. **White, W. F., Nadler, J. V., and Cotman, C. W.**, The effect of acidic amino acid antagonists on synaptic transmission in the hippocampal formation *in vitro*, *Brain Res.*, 164, 177, 1979.
29. **Koerner, J. F. and Cotman, C. W.**, Micromolar L-2 amino-4-phosphonobutyric acid selectively inhibits perforant path synapses from lateral enthorinal cortex, *Brain Res.*, 216, 192, 1981.
30. **Lanthorn, T. H. and Cotman, C. W.**, Baclofen selectivity inhibits excitatory synaptic transmission in the hippocampus, *Brain Res.*, 225, 171, 1981.
31. **Storm-Mathisen, J. and Wold, J. E.**, *In vivo* high affinity uptake and axonal transport of D-[2,3-^{3}H] aspartate in excitatory neurons, *Brain Res.*, 230, 427, 1981.
32. **Gall, C., Brecha, N., Karten, H. J., and Chang, K.-J.**, Localization of enkephalin-like immunoreactivity to identified axonal and neuronal populations of the rat hippocampus, *J. Comp. Neurol.*, 198, 335, 1981.
33. **Stengaard-Pedersen, Fredens, K., and Larsson, L. I.**, Enkephalin and zinc in the hippocampal mossy fibre system, *Brain Res.*, 212, 230, 1981.
34. **Haug, F. M. S.**, On the normal histochemistry of trace metals in the brain, *J. Hirnforschung*, 16, 151, 1975.
35. **Sloviter, R. S.**, A selective loss of hippocampal mossy fiber TIM stain accompanies granule cell seizure activity induced by perforant path stimulation, *Brain Res.*, 330, 150, 1985.
36. **Fonnum, F., Karlsen, R. L., Malthe-Sørenssen, D., Skrede, K. K., and Walaas, I.**, Localization of neurotransmitters, particularly glutamate, in hippocampus, septum, nucleus accumbens and superior colliculus, *Prog. Brain Res.*, 51, 167, 1979.
37. **Wieraszko, A. and Lynch, G. S.**, Stimulation-dependent release of possible transmitter substances from hippocampal slices studies with localized perfusion, *Brain Res.*, 160, 372, 1979.
38. **Spencer, H. J., Tominez, G., and Halpern, B.**, Mass spectrographic analysis of stimulated release of endogenous amino acids from rat hippocampal slices, *Brain Res.*, 212, 194, 1981.
39. **Wieraszko, A.**, Stimulation-dependent uptake of glutamic acid by hippocampal slices, *Brain Res.*, 207, 209, 1981.
40. **Storm-Mathisen, J. and Woxen-Opsahl, M.**, Aspartate and/or glutamate may be transmitters in hippocampal efferents to septum and hypothalamus, *Neurosci. Lett.*, 9, 65, 1978.
41. **Zaczek, R., Hedreen, J. C., and Coyle, J. T.**, Evidence for hippocampal-septal glutamergic pathway in the rat, *Exp. Neurol.*, 65, 145, 1979.
42. **Collingridge, G. L., Kehl, J. S., and McLennan, H.**, The antagonism of amino acid induced excitations of rat hippocampal CA1 neurons *in vitro*, *J. Physiol. (London)*, 334, 19, 1982.

43. **Collingridge, G. L., Kehl, J. S., and McLennan, H.,** Excitatory amino acids in synaptic transmission in the Schaffer collateral-commissural pathway of the rat hippocampus, *J. Physiol. (London),* 334, 33, 1983.
44. **Koerner, J. F. and Cotman, C. W.,** Response of Schaffer collateral CA1 pyramidal cell synapses of the hippocampus to the analogues of acidic amino acids, *Brain Res.,* 251, 105, 1982.
45. **Walaas, I. and Fonnum, F.,** Biochemical evidence for glutamate as a transmitter in hippocampal efferents to the basal forebrain and hypothalamus in the rat brain, *Neuroscience,* 5, 1691, 1980.
46. **Malthe-Sørenssen, D., Odden, E., and Walaas, I.,** Selective destruction by kainic acid of neurons innervated by putative glutamergic afferents in septum and nucleus of the diagonal band, *Brain Res.,* 182, 461, 1980.
46a. **Divac, I. and Fonnum, F.,** unpublished.
47. **Meibach, R. and Segal, A.,** Efferent connection of the hippocampal formation in the rat, *Brain Res.,* 124, 197, 1977.
48. **Meeker, R. and Myers, R. D.,** *In vivo* [^{14}C] amino acid profiles on discrete hypothalamic regions during push-pull perfusion in the unrestrained rat, *Neuroscience,* 4, 495, 1978.
49. **Walaas, I. and Fonnum, F.,** The effect of parenteral glutamate treatment on the localization of neurotransmitters in the mediobasal hypothalamus, *Brain Res.,* 155, 549, 1978.
50. **Olney, J. W.,** Brain lesions, obesity and other disturbances in mice treated with monosodium glutamate, *Science,* 164, 719, 1969.
51. **Storm-Mathisen, J., Leknes, A. K., Bore, A. T., Vaulard, J. L., Edminson, P., Haug, F. M. S., and Ottersen, O. P.,** The first visualization of glutamate and GABA in neurons by immunocytochemistry, *Nature,* 301, 812, 1981.
52. **Wenthold, R. J. and Altschuler, R. A.,** Immunocytochemistry of aspartate aminotransferase and glutaminase, in *Glutamine, Glutamate and GABA in the Central Nervous System,* Hertz, L., Kvamme, E., McGeer, E. G., and Schousboe, A., Eds., Alan R. Liss, New York, 1983, 33.
53. **Wolf, G., Schünzel, G., and Storm-Mathisen, J.,** Lesion of Schaffer's collaterals in the rat hippocampus affecting glutamate dehydrogenase and succinate dehydrogenase activity in the stratum radiatum of CA1. A study with special reference to the glutamate transmitter metabolism, *J. Hirnforschung,* 25, 249, 1984.
54. **Wolf, G., Schünzel, G., and Rothe, E.,** Histochemical demonstration of sodium dependent glutamate uptake in brain tissues by glutamate dehydrogenase reaction, *Exp. Brain Res.,* 62, 659, 1986.
55. **Schmidt, W. and Wolf, G.,** Histochemical localization of aspartate aminotransferse activity in the hippocampal formation and in peripheral ganglia of the rat with special reference to the glutamate transmitter metabolism, *J. Hirnforschung,* 25, 508, 1984.
56. **Dolphin, A. C., Errington, M. L., and Bliss, T. V. P.,** Long-term potentiation of the perforant path *in vivo* is associated with increased glutamate release, *Nature (London),* 266, 377, 1982.
57. **Nadler, J. V., Evenson, D. A., and Smith, E. M.,** Evidence from lesion studies for epileptogenic and non-epileptogenic neurotoxic interactions between kainic acid and excitatory innervation, *Brain Res.,* 205, 405, 1981.
58. **Hamberger, A., Chiang, G. H., Sandoval, E., and Cotman, C. W.,** Glutamate as a CNS transmitter, II. Regulation of synthesis in the releasable pool, *Brain Res.,* 168, 531, 1979.
59. **Fonnum, F., Storm-Mathisen, J., and Divac, I.,** Biochemical evidence for glutamate as neurotransmitter in corticostriatal and corticothalamic fibres in rat brain, *Neuroscience,* 6, 863, 1981.
60. **Fonnum, F., Søreide, A., Kvale, I., Walker, J., and Walaas, I.,** Glutamate in cortical fibers, *Adv. Biochem. Psychopharmacol.,* 27, 29, 1981.
61. **Webster, K. E.,** Cortico-striatal interrelations in the albino rat, *J. Comp. Anat. (London),* 95, 532, 1961.
62. **Goldmann, P. S. and Nauta, W. J. H.,** An intricate patterned prefronto-caudate projection in the rhesus monkey, *J. Comp. Neurol.,* 171, 369, 1977.
63. **Kemp, J. M. and Powell, T. P. S.,** The cortico-striate projection in the monkey, *Brain,* 93, 525, 1970.
64. **Hedreen, J. C.,** Corticostriatal cells identified by the peroxidase method, *Neurosci. Lett.,* 4, 1, 1977.
65. **Divac, I., Fonnum, F., and Storm-Mathisen, J.,** High affinity uptake of glutamate in terminals of corticostriatal axons, *Nature (London),* 266, 377, 1977.
66. **Walaas, I.,** Biochemical evidence for overlapping neurocortical and allocortical glutamate projections to the nucleus accumbens and rostral caudatoputamen in the rat brain, *Neuroscience,* 6, 399, 1981.
67. **Engelsen, B., Westeberg, E., Fonnum, F., and Wieloch, T.,** The effect of insulin-induced hypoglycemia on the concentration of glutamate and related amino acids and energy metabolites in the intact and decorticated rat neostriatum, *J. Neurochem.,* 47, 1634, 1986.
68. **Hassler, R., Haug, P., Nitsch, C., Kim, J. S., and Paik, K.,** Effect of motor and promotor cortex ablation on concentrations of amino acids, monoamines, and acetylcholine and on the ultrastructure in rat striatum. A confirmation of glutamate as the specific corticostriatal transmitter, *J. Neurochem.,* 38, 1087, 1982.

69. **Sandberg, M., Ward, H. K., and Bradford, H. F.**, Effect of corticostriate pathways lesion on the activities of enzymes involved in synthesis and metabolism of amino acid neurotransmitters in the striatum, *J. Neurochem.*, 44, 42, 1985.
70. **Streit, P.**, Selective retrograde labeling indicating the transmitter of neuronal pathways, *J. Comp. Neurol.*, 191, 429, 1980.
71. **Arqueros, L., Abarca, J., and Bustos, G.**, Release of D- ^{3}H aspartic acid from the rat striatum. Effect of veratridine-evoked depolarization, fronto-parietal cortex ablation, and striatal lesions with kainic acid, *Biochem. Pharmacol.*, 34, 1217, 1985.
72. **Fonnum, F., Paulsen, R. H., Fosse, V. M., and Engelsen, B.**, Synthesis and release of amino acid transmitter, in *Excitatory Amino Acids and Epilepsy*, Ben-Ari, Y. and Schwartz, R., Eds., 1986, 285.
73. **Godukhin, O. V., Zharikova, A. D., and Novoselov, V. I.**, The release of labeled L-glutamate acid from rat neostriatum *in vivo* following stimulation of frontal cortex, *Neuroscience*, 5, 2151, 1980.
74. **Reubi, J. C. and Cuenod, M.**, Glutamate release *in vitro* from cortico-striatal terminals, *Brain Res.*, 176, 185, 1979.
75. **Rowlands, G. J. and Roberts, P. J.**, Specific calcium-dependent release of endogenous glutamate from rat striatum is reduced by destruction of the cortico-striatal tract, *Exp. Brain Res.*, 39, 239, 1980a.
76. **Spencer, H. J.**, Antagonism of cortical excitation of striatal neurons by glutamic acid diethylester. Evidence for glutamic acid as an excitatory transmitter in the rat striatum, *Brain Res.*, 102, 91, 1976.
77. **Engelsen, B. and Fonnum, F.**, The effect of methioninesulphoximine, an inhibitor of glutamine synthetase, on the levels of amino acids in the intact and decorticated rat neostriatum, *Brain Res.*, 338, 165, 1985.
78. **Engelsen, B., Westeberg, E., Fonnum, F., and Wieloch, T.**, The effect of insulin-induced hypoglycemia on the concentration of glutamate and related amino acids and energy metabolites in the intact and decorticated rat neostriatum, *J. Neurochem.*, 47, 1634, 1986.
79. **Minchin, M. C. W. and Fonnum, F.**, The metabolism of GABA and other amino acids in rat substantia nigra slices following lesions of the striatonigral pathway, *J. Neurochem.*, 32, 203, 1979.
80. **Korf, J. and Venema, L.**, Amino acids in the substantia nigra in rats with striatal lesions produced by kainic acid, *J. Neurochem.*, 40, 1171, 1981.
81. **Berl, S., Lajhta, A., and Wadsch, H.**, Amino acid and protein metabolism. VI. Cerebral compartments of glutamic acid metabolism, *J. Neurochem.*, 7, 186, 1961.
82. **Paulsen, R. E., Contestabile, A., Villani, L., and Fonnum, F.**, An *in vivo* model for studying brain function of brain tissue temporarily devoided of glial cell metabolism. The use of fluoracetate, *J. Neurochem.*, 48, 1377, 1987.
83. **Nicklas, W. J., Nunez, R., Berl, S., and Duvoisin, R.**, Neuronal glial contributions to transmitter amino acid metabolism: studies with kainic acid-induced lesions of rat striatum, *J. Neurochem.*, 33, 839, 1979.
84. **Wieloch, T.**, Hypoglycemia-induced neuronal damage prevented by an N-methyl-D-aspartate antagonist, *Science*, 230, 681, 1985.
85. **Wieloch, T., Engelsen, B., Westerberg, E., and Auer, R.**, Lesions of the glutamatergic cortico-striatal projections in the rat ameliorate hypoglycemic brain damage in the striatum, *Neurosci. Lett.*, 58, 25, 1985.
86. **Giorgi, D. and Meek, J. L.**, Gamma-aminobutyric acid turnover in rat striatum: effects of glutamate and kainic acid, *J. Neurochem.*, 42, 215, 1984.
87. **Fonnum, F., Iversen, E., and Fosse, V. M.**, The correlation between high affinity uptake of [^{3}H]glutamate and [^{14}C]2-oxoglutarate in brain regions, *Neurosci. Lett.*, submitted.
88. **Dusticier, N., Kerkerion, L., Errami, M., and Nieoullon, A.**, Effects of pyroglutamic acid on corticostriatal glutamatergic transmission, *Neuropharmacology*, 24, 903, 1985.
89. **Wong, P. T. H., McGeer, E. G., and McGeer, P. L.**, Effects of kainic acid injections and cortical lesions on ornithine and aspartate aminotransferases in rat striatum, *J. Neurosci. Res.*, 8, 643, 1982.
90. **Creese, I., Coyle, J. T., Snyder, S. A., and Schwarcz, R.**, Dopaminergic receptors localized on cerebral cortical afferents to rat corpus striatum, *Nature (London)*, 271, 766, 1978.
91. **Martres, M. P., Sokoloff, P., Delandre, M., Schwartz, J. C., Protais, P., and Costentin, J.**, Selection of dopamine antagonists discriminating various behavioral responses and radioligand binding sites, *Naunyn Schmiedebergs Arch. Parmakol.*, 325, 102, 1984.
92. **Rowlands, G. J. and Roberts, P. J.**, Activation of dopamine receptors inhibits calcium-dependent glutamate release from corticostriatal terminals *in vitro*, *Eur. J. Pharmacol.*, 62, 241, 1980b.
93. **Giorgieff, M. F., Kemel, M. L., and Glowinski, J.**, Presynaptic effect of L-glutamic acid on the release of dopamine in rat striatal slices, *Neurosci. Lett.*, 6, 73, 1977.
94. **Mitchell, P. R. and Dogell, N. S.**, Modulation of striatal ^{3}H-glutamic acid release by dopaminergic drugs, *Life Sci.*, 26, 2073, 1980.
95. **Roberts, P. J. and Andersen, S. D.**, Stimulating effect of L-glutamate and related amino acids on [^{3}H] dopamine release from rat striatum, *J. Neurochem.*, 32, 1539, 1979.
96. **Roberts, P. J. and Sharif, N. A.**, Effects of L-glutamate and related amino acids upon the release of ^{3}H dopamine from rat striatal slices, *Brain Res.*, 391, 1978.

97. **Kerwin, R. W., Rupniak, N. M., Jenner, P., and Marsden, C. D.,** A comparison of the effects of acute and one year's continuous neuroleptic treatment on the release of ^{3}H glutamate and ^{3}H acetylcholine from rat striatal slices, *Neuroscience,* 11, 205, 1984.
98. **Kuznetsov, V. I. and Godukhin, O. V.,** Mechanism of methionine enkephalin modulation of glutamatergic transmission in the rat striatum, *Neurosci. Lett.,* 57, 143, 1985.
99. **McGeer, E. G. and McGeer, P. L.,** Duplication of biochemical changes of Huntington's chorea by intrastriatal injections of glutamic and kainic acid, *Nature,* 263, 517, 1976.
100. **McBean, G. J. and Roberts, P. J.,** Chronic infusion of L-glutamate causes neurotoxicity in rat striatum, *Brain Res.,* 290, 372, 1984.
101. **Sladeczek, F., Pin, J. P., Recasens, M., Bockaert, J., and Weiss, S.,** Glutamate stimulates inositol phosphate formation in striatal neurons, *Nature,* 317, 717, 1985.

Chapter 19

GLUTAMINE AND ITS NEUROACTIVE DERIVATIVES IN THE RETINA

Mary J. Voaden

TABLE OF CONTENTS

I. THE RETINA — AN INTRODUCTION

Existing as it does on the inner posterior surface of the eyeball, the retina is the most exposed area of the central nervous system (CNS). It is separated from the external environment only by the transparent tissues of the eye and, therefore receives input of light from the external environment. Its function is to transduce the incoming photons into neuronal signals which give rise, in the brain, to the sensory awareness of our environment termed vision.

Although the retina has evolved to perform a highly specialized task, it remains an easily identifiable component of the CNS and uses most if not all of the neurotransmitters found in other areas of the brain.[1-4] In line with this, it retains much of the metabolic organization associated with synthesis, termination of action, and breakdown of the neuroactive compounds, and, since glutamate, aspartate, and GABA are numbered among its neurotransmitters, it is not surprising that the small compartment glial metabolism, sometimes associated with catabolism of these amino acids, is also evident. The aim of this review is to discuss this metabolism and to consider the locations and functions of glutamate, aspartate, and GABA in the retina. At this stage of our knowledge, care must be taken in translocating the information across species since surprising variations, not only in metabolic organization in retinas but also in the chemical transmitters used by specific cell types, continue to be found.

II. THE CELLS OF THE RETINA

As far as we are aware, all retinas conform to the basic pattern of possessing photoreceptor cells (rods and/or cones) together with five other major neuronal classes: bipolar, horizontal, amacrine, ganglion, and interplexiform cells (Figure 1).[5,6] Biopolars carry visual information centripetally through the tissue from photoreceptor to the ganglion cells, whose axons form the optic nerve, whereas horizontal and amacrine interneurons mediate modulating, lateral interactions in the outer and inner synaptic layers, respectively. Interplexiform cells receive synaptic input from amacrine cells in the inner synaptic layer and themselves synapse onto horizontal or bipolar cells in the outer synaptic layer. They have the potential, therefore, to transmit information centrifugally forming a local feedback loop within the tissue. Conversely, a few may receive synaptic input from horizontal cells, potentially providing a centripetal route that does not involve bipolar cells.[7] Conventional astroglia and microglia are present,[8] but by far the major glial component is the large, radially orientated cells, sometimes termed Müller "fibers" (Figure 1). These not only envelop and separate the neurons, but also, when vascularization is intraretinal, encapsulate the vessels. Metabolically and immunologically, they show properties characteristic of both astroglia and oligodendroglia.[8]

Energy metabolism in Müller cells is strongly related to the extent and type of the retinal blood supply. Accordingly, mitochondria may cluster in the proximal or distal regions of the cell, However, although the plasma membrane does show enzyme protein differences between its apical and basal regions,[8-10] asymmetry in the distribution of glutamate, aspartate, or GABA carriers has not been sought/reported. A striking interspecies difference is that whereas Müller cells of mammalian and, certainly in one instance, an elasmobranch fish accumulate exogenously applied GABA, those of the amphibia, birds, and teleost fish so far investigated do not. This is discussed further in Section IV.B.1.

III. POTENTIAL FUNCTIONS OF GLUTAMATE, ASPARTATE, AND GABA IN THE RETINA

A. The Neurotransmitters of Photoreceptor Cells

Photoreceptors transduce light to nervous energy in their outer limbs. In the dark, the

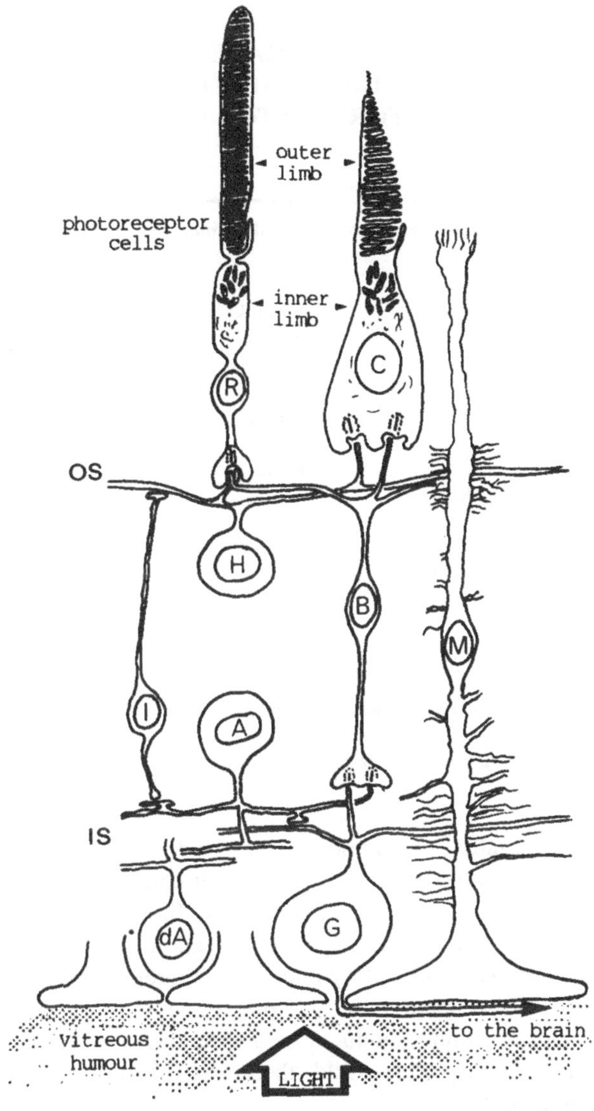

FIGURE 1. Schematic representation of the cells of the retina: R = rod photoreceptors, C = cone photoreceptors, H = horizontal cells, B = bipolar cells, I = interplexiform cells, A = amacrine cells, dA = displaced amacrines, G = ganglion cells, M = Müller glial cells, OS = outer synaptic (plexiform) layer, and IS = inner synaptic (plexiform) layer. It is emphasized that there are many subclasses of retinal neurons. No attempt has been made to show their precise interconnections or their relative sizes.

plasma membrane of an outer limb is freely permeable to sodium ions and a sodium current, generated by Na^+/K^+ ATPase activity in the photoreceptor inner limb, runs extracellularly from the inner to the outer limb; thus, the cell is depolarized. When light is absorbed by visual pigment in the photoreceptor outer limb, the sequence of events that follows leads to closure of the sodium channels and repolarization of the cell.[11] As photoreceptors are depolarized in the dark, they release their neurotransmitters; light stops the release.[17-19]

There is evidence to suggest that glutamate and aspartate are, with species variation, photoreceptor neurotransmitters. The most convincing concerns the action of the amino acids

Table 1
PREDOMINANT SITES OF GLUTAMATE AND ASPARTATE UPTAKE AND OF cAAT-LIKE IMMUNOREACTIVITY IN PHOTORECEPTOR CELLS

		Sites of accumulation		
Species	**cAAT immunoreactivity**	**L-Glutamate**	**L-Aspartate**	**D-Aspartate**
Primates				
Homo sapiens	C (20)	R and C (20) R (32)	R (20,31)	R (20)
Papio papio	—	R and C (30)	R (30)	—
Macaca fascicularis	R (26)	—	—	—
M. nemestrina	C (27)	R (27)	R (27)	—
Cynamolgus iris	—	R (32)	—	—
Cat	—	R (32)	—	—
Guinea pig	C (18,26,28)	—	C (30)	C (29)
Rabbit	—	—	—	C (29)
Rat	C (20) U (18,24,25[a])	R and C (20)	R (20)	—
Goldfish	—	R and C[b] (21)	C[b] (21)	C[b] (21)
Pigeon	—	—	—	C (29)
Toad	—	—	C (33)	C (33)

Note: R = rods, C = cones, and U = unclassified photoreceptors. References are in parentheses.

[a] A large number of cell bodies, suggesting predominantly rods.
[b] Red and green sensitive only.

on bipolar cells. Functionally, two major classes of bipolars are recognized: those that depolarize following light stimulation of photoreceptor cells (''on'' bipolars) and those that are depolarized in the dark and, therefore, respond in light with hyperpolarization (''off'' bipolars). Both ''dark'' responses, i.e., hyperpolarization of ''on'' bipolars and depolarization of ''off'' cells, can be simulated by microiontophoretic application of glutamate and/or aspartate onto the light-adapted retina — the receptors of ''on'' bipolars responding usually with a decrease in sodium conductance and conversely ''off'' bipolars showing an increase.[12-14] Glutamate and aspartate could, therefore, fulfill roles as neurotransmitters, at least for some photoreceptor cells.

Additional electrophysiological evidence comes from the effects of these amino acids on the other neurons postsynaptic to photoreceptors, horizontal cells. Horizontal interneurons are depolarized, not only by the indigenous photoreceptor neurotransmitter(s), but also by application of glutamate or aspartate. Specificity is seen and is species dependent.[2,3,13-16]

With the backing of the electrophysiological studies outlined above, it is of considerable interest to ask how well proposed anatomical markers for Glu/Asp neurons also delineate photoreceptors as aspartatergic or glutamatergic cells. Two major markers are recognized: the cells may possess high-affinity carrier systems for reuptake of the neurotransmitters after release and/or they may contain a higher than average level of the soluble (cytoplasmic) enzyme, aspartate-2, oxoglutarate aminotransferase (glutamate-oxaloacetate transaminase, cAAT). It is suggested that the latter is required for synthesis of the amino acids in the quantities needed for transmission.[17,18] Neither appears to be a totally reliable marker,[18,19] but, with the caveats given at the end of this section, positive results can be considered as evidence for the use of glutamate or aspartate in neurotransmission.

It is clear from the observations, summarized in Table 1, that some photoreceptor cells do indeed possess high-affinity mechanisms that can be used for Glu/Asp accumulation.

Moreover, occasionally the transport systems are seen to have a well-defined preference for one or other of the amino acids. For example, in man, cones are more effective in accumulating L-glutamate than L- or D-aspartate,[20] and in the goldfish it is the rods that show preference for glutamate.[21] Selectivity in the neurotoxicity of kainic acid has suggested that goldfish rods might be glutamatergic,[22] but functional studies, based on the release of GABA from second-order horizontal cells, strongly suggest that the red cones as well, use glutamate rather than aspartate as their neurotransmitter.[23]

Alongside the presence of membrane carriers for Glu/Asp, some photoreceptors also contain high levels of cAAT (Table 1). However, there is variable reaction with the individual antibodies and more studies are needed to clarify the information. Subforms of cAAT exist in the CNS[34] and this may explain the differing immunoreactivities.

In line with the above observations suggesting Glu/Asp as photoreceptor neurotransmitters, there is active synthesis of glutamate and aspartate from exogenously applied, radiolabeled glucose and glutamine in the photoreceptor cell layer of the rat retina[35] and turnover of the "pools" derived from glutamine is reduced on light stimulation as compared with dark adaptation (Section V).[36] Direct evidence for a decreased output of Glu/Asp from light-stimulated photoreceptors has not yet been obtained, although decreased output of aspartate and glutamate in light has been noted with intact rabbit and mudpuppy retinas, respectively.[37,38] Moreover, as 2-amino-4-phosphonobutyric acid and cis-2,3-piperidine dicarboxylic acid were used in the latter study and these are known to block postsynaptic receptors of "on" and "off" bipolars, respectively, it is probable that the result in the mudpuppy does reflect photoreceptor activity.

In summary, and based on current premises, there is convincing evidence that some photoreceptors use glutamate or aspartate as their neurotransmitter. However, it should be noted that photoreceptors contain high levels of the sulfonic amino acid taurine[39,40] and there is the suggestion that this also may have differential effects on "off" and "on" bipolar cells.[39] More recently, immunohistochemical staining for cysteine sulfinic acid decarboxylase (CSAD), a key enzyme in the synthesis of taurine from cysteine, has shown it to be present in the synaptic vesicles of *some* rod and cone photoreceptors of the rat retina,[41] and cysteine is readily taken up and metabolized by photoreceptor cells.[40] Moreover, there is evidence to suggest that cAAT and cysteine sulfinic acid aminotransferase have co-identity.[42,43] Thus, transmitter roles for taurine and its neuroactive relatives cysteine sulfinic and cysteic acids cannot be excluded; more comparative studies are needed.

B. Glutamate, Aspartate, and GABA in the Higher-Order Retinal Neurons

Glutamate and aspartate may also serve as neurotransmitters for some bipolar, amacrine, and ganglion cells of the retina. In addition, there can be little doubt that GABA is a major "inhibitory" retinal neurotransmitter; GABAergic neurons include subpopulations of amacrine cells and, with species variation, horizontal, bipolar, and interplexiform cells.

1. Glutamate and Aspartate

The results from many neurophysiological studies[44] and the widespread neurodegenerative effects of glutamate and kainate in the retina[45] have, for many years, suggested the existence of numerous receptors for excitatory amino acids on higher-order retinal neurons. However, direct demonstration of glutamatergic or aspartatergic cells has proved to be a challenge. In particular, autoradiography, following accumulation of radiolabeled glutamate or aspartate has failed, in general, to delineate specific neurons,[29] perhaps because of masking by the extensive glial labeling that frequently occurs[20,29-31] or because high-affinity carriers are not present in the plasma membrane of the cells.[18] It is of considerable interest, therefore, that immunohistochemical localization of cAAT has shown not only selective staining of photoreceptors (Table 1) but also increased enzyme levels in higher-order retinal neurons. In the cat retina, cAAT immunoreactivity has been described as present in almost every cell body in the inner nuclear layer.[19] Others have been more specific and have identified amacrine (man,[20] monkey,[26] rat,[20,24] and guinea pig[26]), bipolar (man,[20] monkey,[26] and rat[20]), ganglion (monkey,[26] and guinea pig[26]), and horizontal (rat[24]) cells; glial staining has also been seen by some[19,20] but not by others.[24]

As in photoreceptors (Table 1), different antibodies suggest different localizations and there are anomalies to be investigated. Moreover, some enzyme activity may be supplying precursor glutamate for GABA formation in GABAergic neurons. Based on personal observation, this has been proposed by Brandon and Lam,[20] but discounted by Lin and coworkers,[25] and, if present, would account for only a subpopulation of immunoreactive amacrine cells. Overall, therefore, initial results on the distribution of cAAT-like immunoreactivity in the retina add substance to the notion that Glu/Asp are major retinal neurotransmitters — a conclusion strongly supported by the results now being obtained in studies combining intracellular recording with microiontophoretic application of the excitatory amino acids and their agonists and antagonists.[46-49] In addition to roles in intraretinal neurotransmission, there is a substantial body of evidence to suggest that Glu/Asp are ganglion cell transmitters in some species[2,26,29,50-53] but not in others.[54,55]

The synthesis and turnover of glutamate and aspartate in the inner retinal layers in light and dark are discussed in Section V.

2. *GABA*

It has been apparent from the earliest studies that there is selective, high-affinity uptake of GABA into certain retinal neurons as well as, in some species, entry in glia. With the advent of immunohistochemical techniques, glutamic acid decarboxylase (GAD) and endogenous GABA have also been localized and we can now compare the results obtained (Table 2).

Based on the criteria of possessing GAD-like immunoreactivity and endogenous GABA, all the species studied appear to have subpopulations of GABAergic amacrine cells. In general, these are found on both borders of the inner synaptic layer. The ones lying proximally, intermingled with ganglion cells, are known as displaced amacrines (Figure 1). There is conflicting data regarding the association of GABA with true ganglion cells as Ehinger and Falck[82] and Yazulla[4] have found positive identification, whereas Brandon[59] and Osborne et al.[75] have obtained negative results. Some mammals, e.g., man, the cat, and rat, may have GABAergic interplexiform cells, the parikarya lying with distal amacrines, and, in birds, teleost fish, amphibia, and reptiles, subpopulations of horizontal cells are likely to employ GABA for neurotransmission. Antibodies raised against GABA-protein complexes and appearing highly specific for GABA[4,75] also show a variable reactivity with some mammalian horizontal cells (Table 2). The significance is unknown; there is no evidence to suggest that the cells will accumulate exogenously applied GABA and they do not contain GAD at levels detectable by immunohistochemistry. Nonetheless, cAAT[24] and GABA-α-oxoglutarate aminotransferase (GABA-T)[25] may be present.

At first sight it appears bizarre that a few bipolar cells in the retina of frogs and some other amphibia contain GABA and GAD and, therefore, may employ GABA as a neurotransmitter. However, glycine, the other major "inhibitory" neurotransmitter in the retina is also associated with specific bipolar cells (including some of man).[83] We do not know functions.

An impression derived from Table 2 is that, in general, there is good correspondence between the sites of uptake of exogenously applied GABA and the endogenous locations of GAD and GABA. To an extent this is true, particularly for horizontal, interplexiform, and, in amphibia, bipolar cells. However, in birds and fish, the pattern of GABA uptake does not always correspond with GAD-like immunoreactivity nor with postsynaptic GABA bind-

Table 2
LOCALIZATION OF ENDOGENEOUS GABA (G), GAD-LIKE IMMUNOREACTIVITY (I), AND SITES OF GABA UPTAKE IN VITRO (U) IN THE VERTEBRATE RETINA

Species		Cells of the retina[a]						
		H	B	Ip	A	dA	M	Ref.
Man	G			+	+	+		4, 56
	I			+	+	+		4, 56
	U				+?		+	31, 32
Monkey	G			+	+	+		56, 57
Baboon	I			±	+	+		56, 57, 59
	U				+?		+	32, 70
Cat	G	±		+	+	+		4, 75
	I			+	+	+		59
	U			+	+	+?	+	32, 63, 66, 70, 73
Rabbit	G	±		+	+	+		4, 75
	I				+	+		58, 59
	U				+?		+	32, 60, 69, 70
Rat	G	±		+	+	+		4, 75
	I			±	+	+		25, 59, 68
	U				+?		+	32, 65, 72
Pigeon	G	+			+	+		4, 75
Chicken	I	+			+			59, 76
	U	+			+	+		4, 71, 79
Goldfish	G	+			+	+		4, 75
Catfish	I	+			+	+		4, 59, 62
	U	+			+			61, 64, 78
Skate	I	+			+	+		4
	U				+?		+	8, 74, 80
Frog	G	+	+		+	±		4, 75
Xenopus	I	+	+		+	±		4, 58
Mudpuppy	U	+	+		+	±		4, 67, 77, 81
Turtle	G	+			+			4
	I	+			+			4
	U	+			+			80

Note: + = Positive, ± = conflicting observations, ? = doubtful, H = horizontal cells, B = bipolar cells, Ip = interplexiform cells, A = amacrine cells, dA = displaced amacrines present in the ganglion cell layer; and M = Müller cells.

[a] It is emphasized that only subpopulations of these neurons are potentially GABAergic.

ing.[4] In the rabbit retina as well, there are some amacrine cells that apparently lack GAD, but possess a high affinity uptake mechanism that will accept GABA.[4] The significance remains to be determined.

In line with the diverse locations in which GABA is found, functions are numerous. In lower-order species, where it is employed by horizontal cells, it provides feedback modulation of photoreceptor activity, thereby heightening contrast features in the visual image, and potentially is involved in color processing.[4,84] When released from interplexiform cells, it will, in part, modulate horizontal and/or bipolar cell function and in amacrine cells it is playing a central role in forming and modulating the complex functional properties of ganglion cells such as movement detection and directional and orientation specifity,[4,84] exerting some of its actions via cholinergic interneurons.[85] GABA is also known to modulate, both directly and indirectly, the activity of retinal dopaminergic cells *(vide infra),* perhaps, in part, affecting the higher order processes concerned with light adaptation.[1,2,4,86]

Since GABA is subserving many functions in retinas, complex changes in turnover might be expected at various levels of light stimulation. In addition, species variation is predictable.

A number of observations suggest synaptic release of GABA in the dark-adapted rat retina.[87-89] In particular, there is evidence that GABAergic amacrine cells synapse directly onto dopaminergic neurons and sustain a tonic inhibition in the dark.[86,88] Exposure to light increases the rate of dopamine biosynthesis and release — a response compatible with a decrease in GABA turnover *(vide infra).* However, the binding of GABA and benzodiazepine to their respective receptors also decreases,[89] potentially contributing to the effect, and there is modulation of the system by melatonin, released from dark-adapted photoreceptors.[90] As well as providing a direct tonic inhibition in the dark, another subpopulation of GABAergic neurons may be involved in a pathway that augments tyrosine hydroxylase activity in the light-adapted tissue.[86,88]

The initial rate of GABA accumulation, following inhibition of its catabolic enzyme GABA-T, with, e.g., γ-vinyl GABA or gabaculine, potentially provides an index of GABA utilization.[91] So far, in studies on the retina, this method has been applied only to the chick[92] and rat,[87,93] and results with the rat suggest that the total turnover of GABA decreases when the tissue is exposed to low levels of illumination.[93] Rates may rise again at higher light levels,[87,93] but there are anomalies[93] and more work is needed. Label and chase techniques, using ^{3}H-glutamine as precursor, add some support to these findings as the relative specific activity of GABA in the isolated, dark-adapted rat retina decreases on exposure to dim light (Section V) and may rise again in bleached tissue.[94]

In marked contrast to the situation in the rat retina, GABA turnover and concentration are significantly increased in goldfish and frog tissue in the light.[95-98] It is known that horizontal cells are involved in goldfish,[99,100] but the changes in the frog retina need to be more precisely localized than heretofore.[97] Apparent similarities between the species may be superficial as autoradiography has shown that whereas exposed grains are present over horizontal cells of both light- and dark-adapted frog retina, preloaded with ^{3}H-GABA,[67] it is only cells of the light-adapted goldfish that are clearly labeled.[64] Goldfish horizontal cells possess a voltage-dependent, Ca-independent release mechanism for GABA that is thought to be involved at feedback synapses in photoreceptor terminals. This has been implicated in the light/dark differences in GABA accumulation[4,99] and may not be present in frog horizontal cells.[4,101] In the frog retina, it is labeling of the amacrine cells that is variable, being greatly diminished in the light as compared with the dark.[67] The cause is not known.

IV. COMPARTMENTATION OF AMINO ACID METABOLISM IN THE RETINA

A. Endogenous Localization of Glutamate, Aspartate, and GABA

Aspartate, glutamate, and GABA are individually present in retinas at concentrations of about 2.0 to 5.0 μmol/g wet weight and are principally found in neurons (which in the rat constitute about 85% of the total cell volume).[102] Corresponding with cAAT distribution (Table 1, Section III.A and B.1), glutamate and aspartate are fairly evenly distributed between photoreceptors and the inner retinal layers.[103] In contrast, GABA is considerably more concentrated in the "inner" neurons with peak levels in the inner synaptic layer, where high levels of GAD activity and immunoreactivity (Table 1) have been routinely observed. In the rat retina, glutamate, aspartate, and GABA formation from both glucose and glutamine occur predominantly in locations consistent with the distribution of their anabolic enzymes.[35,36,104]

B. Sites of Uptake

A working premise for many years has been that most, if not all, neuroactive amino acids that are fulfilling a function based on their release from a neuron can be taken up again into the cells releasing them by active, high-affinity mechanisms; examples in the retina for glutamate, aspartate, and GABA have been given in Section III and Tables 1 and 2. However, it is also recognized that glial cells provide potentially major "reservoirs" for removing these compounds from the extracellular space and metabolizing them to nonneuroactive derivatives. In line with this, glial uptake of radiolabeled glutamate, aspartate, and GABA has been observed in many autoradiographical studies. There are, however, striking species differences.

1. GABA

The major variation, emphasized by Table 2, is that the Müller cells of some lower-order vertebrates do not accumulate GABA. We do not know why. However, an obvious pattern that emerges is that there is no glial uptake where there are GABAergic horizontal cells that actively accumulate GABA and show high immunoreactivity for GAD. Future studies will establish the physiological basis for the species difference and if there is a connection with horizontal cell homeostasis. However, apart from the existence of voltage-dependent GABA transport systems, which may not be universally present across species (Section III.B.2), horizontal cells are interconnected by gap junctions which are affected, perhaps indirectly, by the GABA antagonists picrotoxin and bicuculline.[4] They may, therefore, require a special microenvironment.

It is intriguing that immunohistochemical techniques have failed to detect GABA in mammalian Müller cells.[4,57,75] Time will tell if this merely reflects a lack of GABA turnover in the retinas being studied or whether GABA, when taken up, is rapidly metabolized and/or is stored in a nonreactive form. In the rat retina, where exogenously applied GABA is accumulated predominantly, if not solely, by Müller cells (Table 2), the rate of release of label from preloaded ^{14}C(U)-GABA is greatly reduced if GABA metabolism is inhibited.[105]

2. Glutamate and Aspartate

Glutamate and aspartate are taken up by Müller cells including those that do not accumulate GABA. However, as with GABA, relative proficiencies of the carrier(s) for the amino acids need to be evaluated and area differences within a retina need to be systematically studied. For example, autoradiography has suggested that baboon Müller cells will readily accumulate ^{3}H-L-glutamate, but not ^{3}H-L-aspartate,[30] whereas, in other species, ^{3}H-L-aspartate is also taken up proficiently.[20,30,31,33] In addition, Beaudet and co-workers[50] have noted that Müller cells in the central retina of the pigeon are more intensively labeled by ^{3}H-D-aspartate than are those of the periphery.

C. Metabolic Compartmentation

As in other areas of the CNS, glutamine synthetase in the retina is exclusively localized

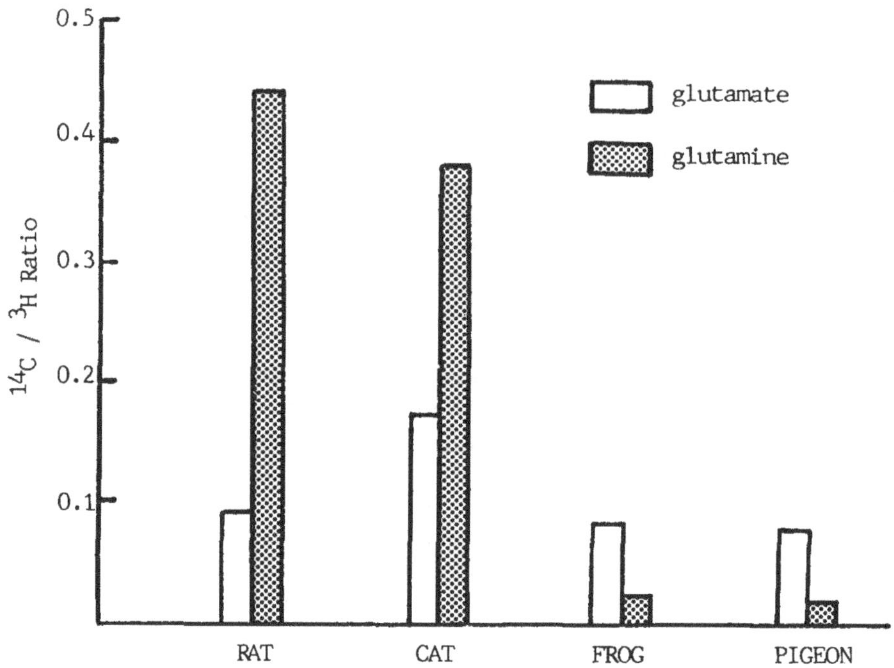

FIGURE 2. The formation of glutamate and glutamine from ^{14}C-GABA in retinas from the rat, cat, frog, and pigeon. Retinas were incubated for 60 min in Krebs' bicarbonate medium (tonicity adjusted to 280 mOsm for the frog) containing 30 μCi/mℓ ^{14}C-GABA (232 mCi/mmol). The incubation temperature was 25°C for the frog and 37°C for the other species. They were then homogenized in 1.0 mℓ of 75% ethanol and the supernatant was evaporated. After reconstitution in 50 μℓ 0.05 *M* sodium bicarbonate, portions were reacted with ^{3}H-dansyl chloride as described in Reference 108. Dansylated amino acids were separated by TLC on micropolyamide plates and the individual spots (viewed under UV light) were recovered. Radioactivity was counted by scintillation spectrometry. Whereas all the tissue free amino acids are dansylated and are labeled with tritium, only those derived from ^{14}C-GABA contain radioactive carbon. The $^{14}C/^{3}H$ ratio of each sample is, therefore, a measure of the specific activity of the amino acid within the tissue. In the rat and cat, the specific activity of glutamine exceeds that of its precursor glutamate, whereas the reverse is true in the frog and pigeon.[108]

to glial cells.[106] Consequently, a major difference following amino acid catabolism in neurons and glia is that it is only in the latter that glutamine is subsequently formed. Moreover, only the glial glutamate pool can serve as the immediate precursor of glutamine. Since the major "pool" of glutamate in the CNS are found in neurons, radiolabeled substrate, metabolized in glia, will produce glutamine which, relative to the whole tissue, has a higher specific activity than that of its precursor — an apparent anomaly. This now classic observation, reviewed by Berl and Clarke,[107] is seen in the retina as well as in other areas of the CNS and provides a simple assessment of whether metabolism is occurring in glia or neurons. Figure 2 gives examples and shows that when ^{3}H-GABA is taken up and metabolized predominantly by Müller cells, as in cat and rat retinas, the resultant glutamine has a higher specific activity than the total tissue glutamate. In contrast, when ^{3}H-GABA is metabolized in frog and pigeon retinas, the ^{3}H-glutamine produced has a much lower specific activity than that of the total glutamate pool. EM and LM autoradiography has implied that no GABA enters Müller cells of the frog and pigeon. However, some label must be entering the glia for radioactive glutamine to be formed. This could be as GABA or as radiolabeled products that are being released from the neurons and, secondarily, taken up and metabolized by Müller cells. Presumably, autoradiography does not reveal the glial label because it is lost during processing and/or is insufficient. Exogenous ^{3}H-glutamate, taken up by frog

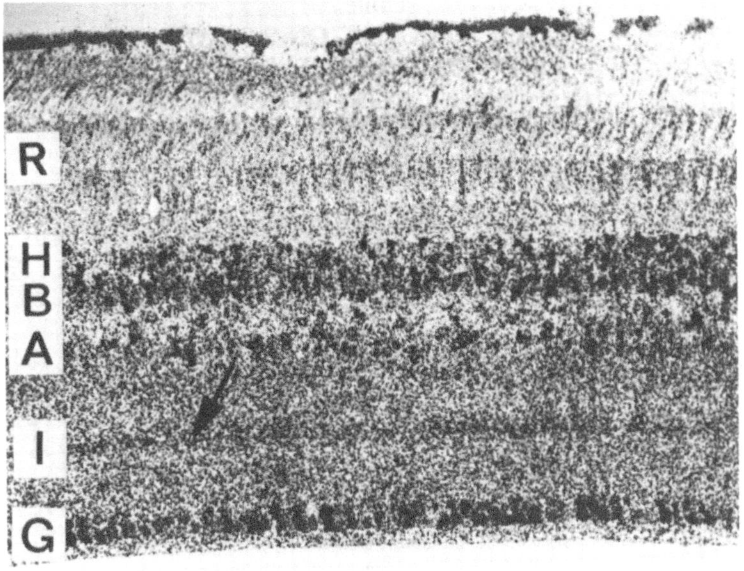

FIGURE 3. Autoradiograph showing sites of uptake of radioactive glutamine in the pigeon retina. Samples from the "red spot" of the pigeon retina were incubated for 30 min at 37°C in Krebs' bicarbonate medium containing 100 μCi/mℓ ^{3}H-glutamine, final concentration 4.35 μ*M*. Cells in positions characteristic of horizontal (H), bipolar (B), amacrine (A), displaced amacrine (G), and ganglion (G) cells are heavily labeled. Exposed grains (arrowed) are also more numerous in the middle of the inner synaptic layer (I); R photoreceptor cells. (× 190.) (After Voaden, M. J., Lake, N., Marshall, J., and Morgaria, B., *J. Neurochem.*, 31, 1069, 1978.)

Müller cells, does yield ^{3}H-glutamine with a higher specific activity than the total tissue ^{3}H-glutamate.[109]

V. GLUTAMINE AND GLUCOSE AS PRECURSORS OF GLUTAMATE, ASPARTATE, AND GABA IN THE RETINA

Since glutamine is continuously being formed from other amino acids that enter the extracellular space and then are cleared by the glia, and since it can also enter the CNS from the blood-stream, it is, potentially, a ubiquitous substrate. In vitro it readily passes out from its site of formation and accumulates in the incubation medium;[110] similarly, in vivo, 200 to 600 μ*M* is maintained in the vitreous body.[36,111,112] Autoradiography has suggested that it will freely enter most cells of the retina[60,104] (Figure 3) and tracer studies have shown that it is metabolized alongside glucose in neurons to form, predominantly, glutamate, aspartate, and GABA.

There is species variation in the relative amounts of the products formed.[104] For example, in the rat retina, incubated for 5 min with ^{3}H-glutamine and then washed for 55 min with medium containing unlabeled substrate, 56% of the total tritium remaining in the retina is found in GABA and 20% is found in glutamate. Corresponding with cAAT and GAD distribution (Section III.B, Table 2), cell bodies located in the amacrine cell layer of the tissue contain a predominant amount of the radioactivity. In contrast, when the same protocol is applied to the pigeon retina, only 18% of the total residual radioactivity is found in GABA, whereas 67% is present in glutamate. In this instance, cell bodies in the bipolar, amacrine, and ganglion cell layers of the tissue are diffusely labeled. Grain density is enhanced when the tissue is incubated for a longer period with ^{3}H-glutamine (Figure 3), the label present

Table 3
THE INCORPORATION OF LABEL FROM ^{14}C-GLUCOSE AND ^{14}C-GLUTAMINE INTO AMINO ACIDS IN THE ISOLATED, DARK-ADAPTED RAT RETINA

	Relative specific activity[a] ($\times 10^2$)	
	^{14}C-glucose + glutamine	^{14}C-glutamine + glucose
Aspartate	8.1	7.2
Glutamate	19.1	17.9
GABA	4.6	8.3

[a] Specific activity of the amino acid relative to that of the precursor at the beginning of the incubation. For ^{14}C-glucose this was 11.8×10^{-3} dpm/nmol and for ^{14}C-glutamine it was 55.5×10^{-3} dpm/nmol. Glucose was present in both media at 5.5 m*M* and glutamine was present at 600 μ*M*. Retinas from dark-adapted animals were incubated in the dark for 30 min at 37°C with the labeled precursors. The labeled products were then isolated as described in Reference 36 and the radioactivity was counted. The concentration of endogenous amino acids was estimated by double label dansylation in retinas incubated as above but with nonradioactive substrates.

in the middle portion of the inner synaptic layer being noteworthy as it is a feature also seen after application to the pigeon retina of ^3H-D-aspartate.[29,50] Thus, it may represent glutamatergic synaptic terminals.

Comparative studies on the relative involvement of glucose and glutamine in retinal amino acid metabolism have been attempted only with rat tissue. At the levels found in vivo, both molecules potentially contribute proficiently to the carbon chains of the neuroactive amino acids (Table 3). Moreover, as the presence of glutamine has no effect on the entry of carbon atoms derived from glucose,[35] it would appear that the pathways have a measure of independence. This conclusion is supported by the observation that exposure of the dark-adapted rat retina to low-level illumination significantly decreases the labeling of amino acids derived from radioactive glutamine, but has little effect on those formed from glucose (Table 4). Additional studies on tangentially bisected tissue have confirmed that the specific activities of glutamate and aspartate in both the photoreceptor cell and inner retinal layers and of GABA in the inner retina are decreased,[36] but how much the changes reflect synaptic as compared with cell body functioning remains to be evaluated. Nevertheless, the findings do suggest that the pools of the amino acids derived from glutamine are more directly involved in neuronal function than are those served by glucose. In the brain, glutamine is known to be a major and preferential precursor of the synaptically active pools of the neuroactive amino acids,[113-115] and it would not be surprising if the same were true for the retina.

Decreased specific activities might reflect decreased turnover, but they might also arise because increased turnover has led to the release of newly formed, more highly labeled pools, not equilibrated with the general tissue stores. This is unlikely in rods (which constitute about 95% of rat photoreceptors)[116] as they release their neurotransmitter(s) in the dark (Section III.A). However, we do not know the functional characteristics of the putative Glu/

Table 4
THE METABOLISM OF RADIOLABELED GLUCOSE AND GLUTAMINE BY LIGHT-STIMULATED AND DARK-ADAPTED RAT RETINAS IN VITRO

	Substrates			
	^{14}C-glucose + glutamine		^{14}C-glutamine + glucose	
Products	**Dark**	**Light**	**Dark**	**Light**
Aspartate	25 ± 3	23 ± 0.3	62 ± 3	32 ± 3[a]
Glutamate	99 ± 10	99 ± 3	394 ± 10	273 ± 5[a]
Glutamine	14 ± 1	12 ± 0.3	720 ± 21	700 ± 18
GABA	18 ± 2	15 ± 1	157 ± 13	107 ± 3[b]
Total tissue radioactivity	669 ± 16	389 ± 23[a]	1379 ± 15	1145 ± 13[a]

Note: Results are expressed as the mean ±SEM (n = 4) and represent dpm × 10^{-3}/mg protein. Retinas from rats previously dark-adapted for 48 hr were dissected out under dim red light and incubated either in darkness or in ambient daylight attenuated to about 5 × 10^{-2} lux: the aim was to be within the working range of rod photoreceptor cells.[36] Incubations were at 37°C for 30 min in Krebs' bicarbonate medium containing 5.5 m*M* glucose and 600 μ*M* glutamine, together with 30 μCi/mℓ [U-^{14}C]-glucose or 15 μCi/mℓ L-[4-^{14}C]glutamine. The labeled products were then isolated as described in Reference 36 and the radioactivity was counted.

[a] $p < 0.0001$.
[b] $p < 0.05$.

Asp amacrine, bipolar, and horizontal cells detected in the rat retina by cAAT immunoreactivity[20,24] (Section III.B.1) and, therefore, do not know if the changes reflect decreased or increased output.

Evaluation of aspartate and glutamate metabolism is complicated also by the fact that glutamate is the immediate precursor of GABA and, in addition to its own functions, will be affected by changes in GABA utilization. Overall, GABA turnover appears to be attenuated when dark-adapted rats are stimulated with dim light (Section III.B.2). In line with this, the specific activity of GABA formed from glutamine was reduced in the in vitro studies, outlined in Table 4.[36]

The above results provide preliminary evidence suggesting that, in the rat retina, the carbon atoms of glutamine rather than those of glucose are the preferred source for manufacturing the aspartate, glutamate, and/or GABA used in neurotransmission. Moreover, they indirectly add support to the premise that a glial-neuronal glutamine "cycle" is active in the retina. In this scheme, neuroactive amino acids, released as transmitters from nerve endings, are taken up by glial cells, while glutamine, released from the glia, is available to resynthesize the relevant neurotransmitters in the neurons — a concept that would explain why, in mammals, the relative density of label, seen after intravitreal administration of glutamate, aspartate, or GABA, shifts with time from glia to neurons.[32,60,82,117] It must be emphasized, however, that glutamine is not the only compound potentially maintaining the glial/neuronal homeostatic balance. For example, it is recognized that α-ketoglutarate may also be involved[118] and there are several other examples of glial neuronal interchange of carbon and other atoms.[8,106]

We are dealing with a status quo that is achieved because of the flexibility, in use and movement around the tissue, of intermediary linking compounds such as glutamine. However, we are only in the early stages of understanding this balance, discussed in depth by van Gelder,[119] and how it is regulated across species. We know, for example, that in contrast to the situation as it presently appears in the rat[36,95] the pools of neuroactive amino acids, derived from glucose in the frog retina in vivo, are affected by light.[95] Here the relative proficiency of glutamine as a precursor for the light-sensitive pools has not been studied. Although, as the experiment cited was done in vivo, glutamine would have been available to compete with the administered ^{14}C-glucose.

Clearly there are many more studies to be done, and in these we shall have to consider not only species differences, but also both rod- and cone-mediated pathways and how the level and type of light stimulus (flashing vs. steady) is influencing our results.[120] Whatever the outcome, it is predictable that glutamine will be found to be playing a central role in the maintenance of retinal homeostasis and function.

REFERENCES

1. **Graham, L. T., Jr.**, Comparative aspects of neurotransmitters in the retina, *Eye*, 6, 283, 1974.
2. **Ehinger, B.**, Neurotransmitter systems in the retina, *Retina*, 2, 305, 1982.
3. **Brecha, N.**, Retinal neurotransmitters: histochemical and biochemical studies, in *Chemical Neuroanatomy*, Emson, P. C., Ed., Raven Press, New York, 1983, 85.
4. **Yazulla, S.**, GABAergic mechanisms in the retina, *Prog. Retinal Res.*, 5, 1, 1986.
5. **Dowling, J. E., Ehinger, B., and Hedden, W. L.**, The interplexiform cell: a new type of retinal neuron, *Invest. Ophthalmol.*, 15, 916, 1976.
6. **Walls, G. L.**, *The Vertebrate Eye*, Cranbrook Institute of Science, Bloomfield Hills, Mich., 1972.
7. **Marc, R. E. and Liu, W.-L. S.**, Horizontal cell synapses onto glycine-accumulating interplexiform cells, *Nature (London)*, 312, 266, 1984.
8. **Ripps, H. and Witkovsky, P.**, Neuron glia interaction in the brain and retina, *Prog. Retinal Res.*, 4, 181, 1985.
9. **Newman, E. A.**, Regional specialization of retinal glial cell membrane, *Nature (London)*, 309, 155, 1984.
10. **Reichenbach, A., Dettmer, D., Brückner, G., Neumann, M., and Birkenmeyer, G.**, Morphological variability, lectin binding and Na,K^+-activated adenosine triphosphatase activity of isolated Müller (glial) cells from the rabbit retina, *Neurosci. Lett.*, 55, 24, 1985.
11. **Korenbrot, J. I.**, Signal mechanisms of phototransduction in retinal rod, *Crit. Rev. Biochem.*, 17, 223, 1985.
12. **Slaughter, M. M. and Miller, R. F.**, Characterization of an extended glutamate receptor of the bipolar neuron in the vertebrate retina, *J. Neurosci.*, 5, 224, 1985.
13. **Lasater, E. M., Dowling, J. E., and Ripps, H.**, Pharmacological properties of isolated horizontal and bipolar cells from the skate retina, *J. Neurosci.*, 4, 1966, 1984.
14. **Bloomfield, S. A. and Dowling, J. E.**, Roles of aspartate and glutamate in synaptic transmission in rabbit retina, I. Outer plexiform layer, *J. Neurophysiol.*, 53, 699, 1985.
15. **Ishida, A. T.**, Responses of solitary horizontal cells to L-glutamate and kainic acid are antagonized by D-aspartate, *Brain Res.*, 298, 25, 1985.
16. **Slaughter, M. M. and Miller, R. F.**, Identification of a distinct synaptic glutamate receptor on horizontal cells in mudpuppy retina, *Nature (London)*, 314, 96, 1985.
17. **Altschuler, R. A., Neises, G. R., Harmison, G. G., Wenthold, R. J., and Fex, J.**, Immunocytochemical localization of aspartate aminotransferase immunoreactivity in the cochlear nucleus of the guinea pig, *Proc. Natl. Acad. Sci. U.S.A.*, 78, 6553, 1981.
18. **Wenthold, R. J. and Altschuler, R. A.**, Immunocytochemistry of aspartate aminotransferase and glutaminase, *Neurol. Neurobiol.*, 7, 33, 1983.
19. **Bolz, J., Thier, P., and Brecha, N.**, Localization of aspartate aminotransferase and cytochrome oxidase in the cat retina, *Neurosci. Lett.*, 53, 315, 1985.
20. **Brandon, C. and Lam, D. M.-K.**, L-Glutamic acid: a neurotransmitter candidate for cone photoreceptors in human and rat retinas, *Proc. Natl. Acad. Sci. U.S.A.*, 80, 5117, 1983.
21. **Marc, R. E. and Lam, D. M.-K.**, Uptake of aspartic and glutamic acid by photoreceptors in goldfish retina, *Proc. Natl. Acad. Sci. U.S.A.*, 78, 7185, 1981.

22. **Yazulla, S. and Kleinschmidt, J.**, The effects of intraocular injection of kainic acid on the synaptic organization of the goldfish retina, *Brain Res.*, 182, 287, 1980.
23. **Yazulla, S.**, Evoked efflux of [^{3}H]-GABA from goldfish retina in the dark, *Brain Res.*, 325, 171, 1985.
24. **Inagaki, N., Kamisaki, Y., Kiyama, H., Horio, Y., Tohyama, M., and Wada, H.**, Immunocytochemical localizations of cytosolic and mitochondrial glutamic oxaloacetic transaminase isozymes in rat retina as markers for the glutamate-aspartate neuronal system, *Brain Res.*, 325, 336, 1985.
25. **Lin, C. T., Li, H.-Z., and Wu, J.-Y.**, Immunocytochemical localization of L-glutamate decarboxylase, gamma-aminobutyric acid transaminase, cysteine sulfinic acid decarboxylase, aspartate aminotransferase and somatostatin in rat retina, *Brain Res.*, 270, 273, 1983.
26. **Mosinger, J. L. and Altschuler, R. A.**, Aspartate aminotransferase-like immunoreactivity in the guinea pig and monkey retinas, *J. Comp. Neurol.*, 233, 255, 1985.
27. **Sarthy, P. V., Hendrickson, A. E., and Wu, J.-Y.**, Aspartate and glutamate as putative neurotransmitters for photoreceptors in the monkey retina, *Invest. Ophthalmol. Vis. Sci. Suppl.*, 25, 85, 1985.
28. **Altschuler, R. A., Mosinger, J. L., Harmison, G. G., Parakkal, M. H., and Wenthold, R. J.**, Aspartate aminotransferase-like immunoreactivity as a marker for aspartate/glutamate in guinea pig photoreceptors, *Nature (London)*, 298, 657, 1982.
29. **Ehinger, B.**, [^{3}H]-D-Aspartate accumulation in the retina of pigeon, guinea-pig and rabbit, *Exp. Eye Res.*, 33, 381, 1981.
30. **Voaden, M. J., Marshall, J., and Oraedu, A. C. I.**, The biochemistry of photoreceptor cells: metabolic effects of light stimulation and light damage, *Doc. Ophthalmol. Proc. Ser.*, 25, 107, 1981.
31. **Lam, D. M. K. and Hollyfield, J. G.**, Localization of putative amino acid neurotransmitters in the human retina, *Exp. Eye Res.*, 31, 729, 1980.
32. **Bruun, A. and Ehinger, B.**, Uptake of certain possible neurotransmitters into retinal neurons of some mammals, *Exp. Eye Res.*, 19, 435, 1974.
33. **Miller, A. M. and Schwartz, E. A.**, Evidence for the identification of synaptic transmitters released by photoreceptors of the toad retina, *J. Physiol.*, 334, 325, 1983.
34. **Heydorn, W. E., Creed, G. J., Wada, H., and Jacobowitz, D. M.**, Immunological evidence for existence of two subforms of soluble glutamic oxaloacetic transaminase (sGOT) in human and rat brain, *Neurochem. Int.*, 7, 833, 1985.
35. **Morjaria, B. and Voaden, M. J.**, The formation of glutamate, aspartate and GABA in the rat retina: glucose and glutamine as precursors, *J. Neurochem.*, 33, 541, 1979.
36. **Voaden, M. J. and Morjaria, B.**, The synthesis of neuroactive amino acids from radioactive glucose and glutamine in the rat retina: effects of light stimulation, *J. Neurochem.* 35, 95, 1980.
37. **Neal, M. J. and Massey, S. C.**, The release of acetylcholine and amino acids from the rabbit retina in vivo, *Neurochemistry*, 1, 191, 1980.
38. **Miller, R. F., Slaughter, N. N., and Massey, S. C.**, Light and dark dependent release of glutamate and aspartate in the isolated retina of the mudpuppy, *Soc. Neurosci. Abstr.*, 8, 131, 1982.
39. **Cohen, A. I.**, Retinal organization and function: possible roles for taurine, in *Taurine and Neurological Disorders*, Barbeau, A. and Huxtable, R. J., Eds., Raven Press, New York, 1978, 249.
40. **Voaden, M. J., Oraedu, A. C. I., Marshall, J., and Lake, N.**, Taurine in the retina, in *The Effects of Taurine on Excitable Tissues*, Schaeffer, S. W., Kocsis, J. J., and Baskin, S. I., Eds., Spectrum, New York, 1981, 145.
41. **Lin, C.-T., Song, G.-X., and Wu, J.-Y.**, Ultrastructural demonstration of L-glutamate decarboxylase and cysteine sulfinic acid decarboxylase in rat retina by immunocytochemistry, *Brain Res.*, 331, 71, 1985.
42. **Recasens, M. and Delauney, J. P.**, Immunological properties and immunohistochemical localization of cysteine sulfinate or aspartate aminotransferase isoenzymes in rat CNS, *Brain Res.*, 205, 351, 1981.
43. **Recasens, M., Benezra, R., Basset, P., and Mandel, P.**, Cysteine sulfinate aminotransferase and aspartate aminotransferase isoenzymes of rat brain. Purification, characterization and further evidence for identity, *Biochemistry*, 19, 4583, 1980.
44. **Kaneko, A.**, Physiology of the retina, *Annu. Rev. Neurosci.*, 2, 169, 1979.
45. **Morgan, I. G.**, Kainic acid as a tool in retinal research, *Prog. Retinal Res.*, 2, 249, 1983.
46. **Bloomfield, S. A. and Dowling, J. E.**, Roles of aspartate and glutamate in synaptic transmission in rabbit retina. II. Inner plexiform layer, *J. Neurophysiol.*, 53, 714, 1985.
47. **Ikeda, H. and Sheardown, M. J.**, Aspartate may be an excitatory transmitter mediating visual excitation of "sustained" but not "transient" cells in the cat retina: iontophoretic studies *in vivo*, *Neuroscience*, 7, 25, 1982.
48. **Slaughter, M. M. and Miller, R. F.**, Bipolar cells in the mudpuppy retina use an excitatory amino acid transmitter, *Nature (London)*, 303, 537, 1983.
49. **Kato, S., Teranishi, T., and Negishi, K.**, L-Glutamate depolarizes ON-OFF transient type of amacrine cells in the carp retina: an ionophoretic study, *Brain Res.*, 329, 390, 1985.
50. **Beaudet, A., Burkhalter, A., Reubi, J.-C., and Cuénod, M.**, Selective bidirectional transport of [^{3}H]D-aspartate in the pigeon retinotectal pathway, *Neuroscience*, 6, 2021, 1981.

51. **Bondy, S. C. and Purdy, J. L.,** Putative neurotransmitters of the avian visual pathway, *Brain Res.*, 119, 417, 1977.
52. **Kemp, J. A. and Sillito, A. M.,** The nature of the excitatory transmitter mediating X and Y cell inputs to the cat dorsal lateral geniculate nucleus, *J. Physiol.*, 323, 377, 1982.
53. **Sandberg, M. and Corazzi, L.,** Release of endogenous amino acids from superior colliculus of the rabbit: *in vitro* studies after retinal ablation, *J. Neurochem.*, 40, 917, 1983.
54. **Oswald, R. E. and Freeman, J. A.,** Optic nerve transmitters in lower vertebrate species, *Life Sci.*, 27, 527, 1980.
55. **Schmidt, J. T. and Freeman, J. A.,** Electrophysiologic evidence that retinotectal synaptic transmission in the goldfish is nicotinic cholinergic, *Brain Res.*, 187, 129, 1983.
56. **Hendrickson, A., Ryan, M., Noble, B., and Wu, J.-Y.,** Localization of gamma aminobutryic acid (GABA)-containing neurons in *Macaca* monkey and human retina, *Invest. Ophthalmol. Suppl.*, 26, 95, 1985.
57. **Hendrickson, A., Ryan, M., Noble, B., and Wu, J.-Y.,** Colocalization of [^{3}H]-muscimol and antisera to GABA and glutamic acid decarboxylase within the same neurons in monkey retina, *Brain Res.*, 348, 391, 1985.
58. **Brandon, C., Lam, D. M. K., Su, Y. Y. T., and Wu, J.-Y.,** Immunocytochemical localization of GABA neurons in the rabbit and frog retina, *Brain Res. Bull.*, 5, 21, 1980.
59. **Brandon, C.,** Retinal GABA neurons: localization in vertebrate species using an antiserum to rabbit brain glutamate decarboxylase, *Brain Res.*, 344, 286, 1985.
60. **Ehinger, B.,** Glial and neuronal uptake of GABA, glutamic acid, glutamine and glutathione in the rabbit retina, *Exp. Eye Res.*, 25, 221, 1977.
61. **Marc, R. E., Stell, W. K., Bok, D. B., and Lam, D. M. K.,** GABA-ergic pathways in the goldfish retina, *J. Comp. Neurol.*, 182, 221, 1978.
62. **Lam, D. M. K., Su, Y. Y. T., Swain, L., Marc, R. E., Brandon, C., and Wu, J.-Y.** Immunocytochemical localization of L-glutamic acid decarboxylase in the goldfish retina, *Nature (London)*, 278, 565, 1979.
63. **Nakamura, Y., McGuire, B. A., and Sterling, P.,** Interplexiform cell in cat retina: identification by uptake of γ-[^{3}H] aminobutyric acid and serial reconstruction, *Proc. Natl. Acad. Sci. U.S.A.*, 77, 658, 1980.
64. **Lam, D. M. K. and Steinman, L.,** The uptake of [γ-^{3}H]-aminobutyric acid in the goldfish retina, *Proc. Natl. Acad. Sci. U.S.A.*, 68, 2777, 1971.
65. **Neal, M. J. and Iversen, L. L.,** Autoradiographic localization of ^{3}H-GABA in rat retina, *Nature (London) New Biol.*, 235, 217, 1972.
66. **Pourcho, R. G.,** Uptake of [^{3}H]glycine and [^{3}H]GABA by amacrine cells in the cat retina, *Brain Res.*, 198, 333, 1980.
67. **Voaden, M. J., Marshall, J., and Murani, N.,** The uptake of ^{3}H-γ-aminobutyric acid and ^{3}H-glycine by the isolated retina of the frog, *Brain Res.*, 67, 115, 1974.
68. **Vaughn, J. E., Famiglietti, E. V., Barber, R. P., Saito, K., Roberts, E., and Ribak, C. E.,** GABAergic amacrine cells in rat retina: immunocytochemical identification and synaptic connectivity, *J. Comp. Neurol.*, 197, 113, 1981.
69. **Brandon, C., Lam, D. M. K., and Wu, J.-Y.,** The gamma aminobutyric acid system in rabbit retina: localization by immunocytochemistry and autoradiography, *Proc. Natl. Acad. Sci. U.S.A.*, 76, 3557, 1979.
70. **Marshall, J. and Voaden, M. J.,** Autoradiographic identification of the cells accumulating ^{3}H-γ-aminobutyric acid in mammalian retinae: a species comparison. *Vision Res.*, 15, 459, 1975.
71. **Marshall, J. and Voaden, M. J.,** An autoradiographic study of the cells accumulating ^{3}H-γ-aminobutyric acid in the isolated retinas of pigeons and chickens, *Invest. Ophthalmol.*, 13, 602, 1974.
72. **Marshall, J. and Voaden, M. J.,** An investigation of the cells incorporating [^{3}H]GABA and [^{3}H]glycine in the isolated retina of the rat, *Exp. Eye Res.*, 18, 367, 1974.
73. **Freed, M. A., Nakamura, Y., and Sterling, P.,** Four types of amacrine in the cat retina that accumulate GABA, *J. Comp. Neurol.*, 219, 295, 1983.
74. **Bruun, A., Ehinger, B., and Sytsma, V. M.,** Neurotransmitter localization in the skate retina, *Brain Res.*, 295, 233, 1984.
75. **Osborne, N. N., Patel, S., Beaton, D. W., and Neuhoff, V.,** GABA neurons in retinas of different species and their postnatal development *in situ* and in culture in the rabbit retina, *Cell Tissue Res.*, 243, 117, 1986.
76. **Karten, H. J. and Brecha, N.,** Localization of neuroactive substances in the vertebrate retina: evidence for lamination in the inner plexiform layer, *Vision Res.*, 23, 1197, 1983.
77. **Neal, M. J., Cunningham, J., and Marshall, J.,** The uptake and radioautographical localization in the frog retina of [^{3}H]($\pm$)3-aminocyclohexane carboxylic acid: a selective inhibitor of neuronal GABA transport, *Brain Res.*, 176, 285, 1979.
78. **Lam, D. M. K., Lasater, E. M., and Naka, K.-I.,** Gamma-aminobutyric acid: a neurotransmitter candidate for cone horizontal cells of the catfish retina, *Proc. Natl. Acad. Sci. U.S.A.*, 75, 6310, 1978.

79. **Watt, C. B., Su, Y. Y. T., and Lam, D. M. K.,** Interactions between enkephalin and GABA in avian retina, *Nature (London),* 311, 761, 1984.
80. **Lam, D. M. K.,** Synaptic chemistry of identified cells in the vertebrate retina, *Cold Spring Harbor Symp. Quant. Biol.,* 40, 571, 1975.
81. **Voaden, M. J.,** γ-Aminobutyric acid and glycine as retinal neurotransmitters, in *Transmitters in the Visual Process,* Bonting, S. L., Ed., Pergamon Press, Oxford, 1976, 107.
82. **Ehinger, B. and Falck, B.,** Autoradiography of some suspected neurotransmitter substances: GABA, glycine, glutamic acid, histamine, dopamine and L-dopa, *Brain Res.,* 33, 157, 1971.
83. **Marc, R. E. and Liu, W.-L. S.,** (^{3}H) Glycine-accumulating neurons of the human retina, *J. Comp. Neurol.,* 232, 241, 1985.
84. **Daw, N. W., Ariel, M., and Caldwell, J. H.,** Function of neurotransmitters in the retina, *Retina,* 2, 322, 1982.
85. **Masland, R. H., Mills, J. W., and Cassidy, C.,** The functions of acetylcholine in the rabbit retina, *Proc. R. Soc. London Ser. B,* 223, 121, 1984.
86. **Iuvone, P. M.,** Regulation of retinal dopamine biosynthesis and tyrosine hydroxylase activity by light, *Fed. Proc., Fed. Am. Soc. Exp. Biol.,* 43, 2709, 1984.
87. **Proll, M. A. and Morgan, W. W.,** The use of gabaculine-induced accumulation of GABA for an index of synthesis of GABA in the retina, *Neuropharmacology,* 21, 1251, 1982.
88. **Marshburn, P. B. and Iuvone, P. M.,** The role of GABA in regulation of the dopamine/tyrosine hydroxylase-containing neurons of the rat retina, *Brain Res.,* 214, 335, 1981.
89. **Biggio, G., Guarneri, P., and Corda, M. G.,** Benzodiazepine and GABA receptors in the rat retina: effect of light and dark adaptation, *Brain Res.,* 335, 321, 1985.
90. **Dubocovich, M. L., Lucas, R. C., and Takahashi, J. S.,** Light-dependent regulation of dopamine receptors in mammalian retina, *Brain Res.,* 335, 321, 1985.
91. **Bernasconi, R., Maitre, L., Martin, P., and Raschdorf, F.,** The use of inhibitors of GABA-transaminase for the determination of GABA turnover in mouse brain regions: an evaluation of aminooxyacetic acid and gabaculine, *J. Neurochem.,* 38, 57, 1982.
92. **Rando, R. R., Coburn, J., and Parkinson, D.,** The differential effects of GABA-transaminase inactivation in the chick retina and brain, *J. Neurochem.,* 39, 1147, 1982.
93. **Voaden, M. J., Hussain, A. A., Taj, M., and Oraedu, A. C. I.,** Light and retinal metabolism, *Biochem. Soc. Trans.,* 11, 679, 1983.
94. **Voaden, M. J.,** Glutamate, aspartate and GABA metabolism in the rat retina: compartmentation and effects of light, *Neurol. Neurobiol.,* 7, 261, 1983.
95. **Starr, M. S.,** Effect of light stimulation on the synthesis and release of GABA in rat and frog retinae, *Brain Res.,* 100, 343, 1975.
96. **Starr, M. S.,** Effect of dark adaptation on the GABA system in retina, *Brain Res.,* 59, 331, 1973.
97. **Graham, L. T., Baxter, C. F., and Lolley, R. N.,** *In vivo* influence of light or darkness on the GABA system in the retina of the frog *(Rana pipiens), Brain Res.,* 20, 379, 1970.
98. **Nishimura, C., Ida, S., and Kuriyama, K.,** Alteration in GABA system in frog retina following short light and dark adaptations — a quantitative comparison with retinal taurine, *Brain Res.,* 219, 433, 1981.
99. **Ayoub, G. S. and Lam, D. M. K.,** The content and release of endogenous GABA in isolated horizontal cells of goldfish retina, *Vision Res.,* 25, 1187, 1985.
100. **Lam, D. M. K.,** The biosynthesis and content of gamma-aminobutyric acid in the goldfish retina, *J. Cell Biol.,* 54, 225, 1972.
101. **Cunningham, J. R. and Neal, M. J.,** Effect of excitatory amino acids on γ-aminobutyric acid release from frog horizontal cells, *J. Physiol. (London),* 362, 51, 1985.
102. **Rasmussen, K.-E.,** A morphometric study of the Müller cell cytoplasm in the rat retina, *J. Ultrastruct. Res.,* 39, 413, 1972.
103. **Voaden, M. J.,** The localization and metabolism of neuroactive amino acids in the retina, in *Amino Acids as Neurotransmitters,* Fonnum, F., Ed., Plenum Press, New York, 1978, 257.
104. **Voaden, M. J., Lake, N., Marshall, J., and Morjaria, B.,** The utilization of glutamine by the retina: an autoradiographic and metabolic study, *J. Neurochem.,* 31, 1069, 1978.
105. **Voaden, M. J. and Starr, M. S.,** The efflux of radioactive GABA from rat retina *in vitro, Vision Res.,* 12, 559, 1972.
106. **Moscona, A. A.,** On glutamine synthetase, carbonic anhydrase and Müller glia in the retina, *Prog. Retinal Res.,* 2, 111, 1983.
107. **Berl, S. and Clarke, D. D.,** The metabolic compartmentation concept, *Neurol. Neurobiol.,* 7, 205, 1983.
108. **Voaden, M. J., Lake, N., and Nathwani, B.,** A comparison of γ-aminobutyric acid metabolism in neurones versus glial cells using intact isolated retinae, *J. Neurochem.,* 28, 457, 1977.
109. **Kennedy, A. J., Voaden, M. J., and Marshall, J.,** Glutamate metabolism in the frog retina, *Nature (London),* 252, 50, 1974.

110. **Starr, M. S.,** A comparative study of the utilization of glucose, acetate, glutamine and GABA as precursors of amino acids by retinae of the rat, frog, rabbit and pigeon, *Biochem. Pharmacol.*, 24, 1193, 1975.
111. **Coull, B. M. and Cutler, R. W. P.,** Light-evoked release of endogenous glycine into perfused vitreous of the intact rat eye, *Invest. Ophthalmol. Vis. Sci.*, 17, 682, 1978.
112. **Durham, D. G., Dickinson, J. C., and Hamilton, P. B.,** Ion exchange chromatography of free amino acids in human intraocular fluids, *Clin. Chem.*, 17, 285, 1971.
113. **Ward, H. K., Thanki, C. M., and Bradford, H. F.,** Glutamine and glucose as precursors of transmitter amino acids: *in vivo* studies, *J. Neurochem.*, 40, 855, 1983.
114. **Hamberger, A. C., Chiang, G. H., Nylen, E. S., Scheff, S. W., and Cofman, C. W.,** Glutamate as a CNS transmitter. I. Evaluation of glucose and glutamine as precursors for the synthesis of preferentially released glutamate, *Brain Res.*, 168, 513, 1979.
115. **Reubi, J. C.,** Comparative study of the release of glutamate and GABA newly synthesized from glutamine, in various regions of the central nervous system, *Neuroscience*, 5, 2145, 1980.
116. **Cicerone, C. M., Green, D. G., and Fisher, L. J.,** Cone inputs to ganglion cells in hereditary retinal degeneration, *Science*, 203, 1113, 1979.
117. **Blanks, J. C. and Roffler-Tarlov, S.,** Differential localization of radioactive gamma-aminobutyric acid and muscimol in isolated and *in vivo* mouse retina, *Exp. Eye Res.*, 35, 573, 1982.
118. **Shank, R. P. and Campbell, G. L.,** Metabolic precursors of glutamate and GABA, *Neurol. Neurobiol.*, 7, 355, 1983.
119. **Van Gelder, N. M.,** Metabolic interactions between neurons and astroglia: glutamine synthetase, carbonic anhydrase and water balance, in *Basic Mechanisms of Neuronal Excitability*, Jasper, H. H. and van Gelder, N. M., Eds., Alan R. Liss, New York, 1983, 5.
120. **Voaden, M. J.,** Techniques for neurochemical research on the retina, *Res. Methods Neurochem.*, 6, 3, 1985.

Chapter 20

EXCITATORY AMINO ACID RECEPTORS

Jørgen Drejer and Tage Honoré

TABLE OF CONTENTS

I. INTRODUCTION

Phillis et al.[1,2] showed almost 30 years ago that L-glutamate and L-aspartate induce neuronal depolarization after iontophoretic application in the CNS. A number of acidic amino acids structurally related to L-glutamate have since been shown to exert stimulatory effects comparable to or even more potent than those induced by L-glutamate and L-aspartate. This group of substances, classified as excitatory amino acids (EAAs), includes both endogenous compounds (e.g., L-glutamate, L-aspartate, L-cysteate, and quinolinate), substances isolated from natural sources (e.g., quisqualate, kainate, and ibotenate), and synthetic compounds (e.g., *N*-methyl-D-aspartate (NMDA) and α-amino-3-hydroxy-5-methyl-4-isoxazole propionate, AMPA.[2-15]

The stimulatory action of EAAs is exerted via an interaction with specific receptors in neuronal membranes. Association of EAA with such receptors leads to opening of ionophores which allow the influx of certain cations, in particular Na^+. Several classes of distinct EAA receptors have been characterized electrophysiologically and by the use of receptor binding techniques. L-Glutamate and L-aspartate appear to be active on all receptor types with some degree of regional selectivity.[16,17]

II. CLASSIFICATION OF EXCITATORY AMINO ACID RECEPTORS

A. Neuropharmacological Studies

From the first demonstration in 1959 of a depolarizing action of excitatory amino acids, neuropharmacological studies of EAA responses were for almost 2 decades severely limited by the lack of sufficiently potent and selective EAA antagonists. However, in the late 1970s, primarily in the laboratories of Watkins, McLennan, and Curtis, a number of neuropharmacological studies were undertaken on the cat spinal cord in vivo and on the frog spinal cord in vitro. It was found that D-α-aminoadipate (D-αAA) was a weak but selective antagonist of responses to the EAA NMDA, and glutamate diethylester (GDEE) was found to inhibit responses to quisqualate.[7-9,16-24] At the dose levels used, both compounds did not affect kainate responses.[22-24] On the basis of these experiments, EAA receptors were subdivided into NMDA, quisqualate, and kainate receptors.[7,8,16,17] A major breakthrough in EAA neuropharmacology was obtained around 1980 when Watkins developed a series of ω-phosphono-amino acids such as D-α-amino-5-phosphonovaleric acid (D-APV).[25] This substance and the corresponding heptanoic acid analogue (D-APH) were found to be potent antagonists of responses induced by NMDA in the spinal cord without significantly affecting quisqualate and kainate responses.[12,25-31]

Potent and selective quisqualate and kainate antagonists have not been described up to now, and since it has generally been difficult to distinguish between quisqualate and kainate responses, some authors prefer to distinguish only between APV-sensitive NMDA receptors and APV-insensitive non-NMDA receptors.[12] However, the original classification scheme of NMDA, quisqualate, and kainate receptors, which also seems to agree with results from receptor binding studies, is still generally accepted. Tables 1 and 2 give a list of a series of EAA agonists and antagonists, respectively, with more detailed information on the selectivities on different EAA receptor subtypes.

The most potent NMDA antagonists are the phosphono- compounds of mono- or dicarboxylic acids such as D-APV; D-APH; and CPP, 3(2-carboxy-piperazine-4-yl) propyl-L-phosphonic acid; all developed by Watkins and colleagues.[25-33] GDEE has, as mentioned above, been considered to be a selective quisqualate antagonist. However, the substance which in most preparations is extremely weak is not specific, and since the compound is also rather unstable in aqueous solution, its usefulness as a selective quisqualate antagonist has been seriously questioned.[8,12,16-19] A few compounds such as D-glutamylaminomethyl

Table 1
SELECTED EEAs

Compound	Formula (nonionized form)	Excitation of cat spinal neurons (relative potencies)	Excitation via NMDA receptors (APV sensitive)	Excitation via Non-NMDA receptors	Ref.
L-Aspartate	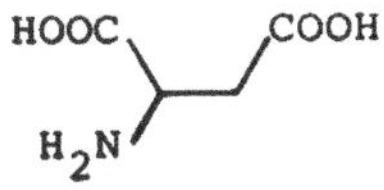	W	+	+	1, 2, 6—8, 20—24
D-Aspartate	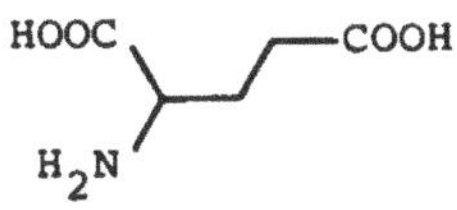	W	+	+	23, 24
L-Glutamate	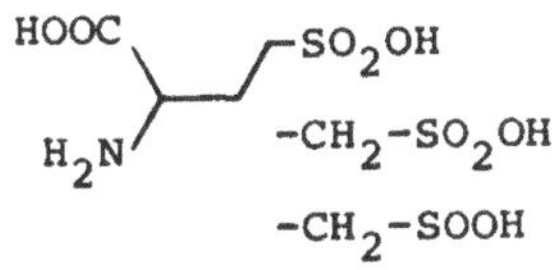	W	(+)	+	1, 2, 6—8, 20—24
D-Glutamate	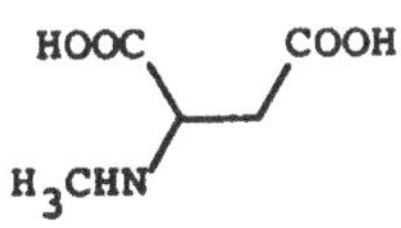	W	+	(+)	22, 23
L-Cysteate		W	+	+	20—24
L-Homocysteate		M	+	+	23
L-Homocysteine sulfinate		M	+	+	23
		M	+	0	6—8, 20—24, 29, 52
Quisqualate	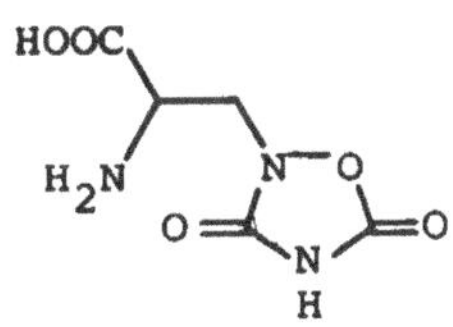	P	0	+	3—8, 20—24
AMPA	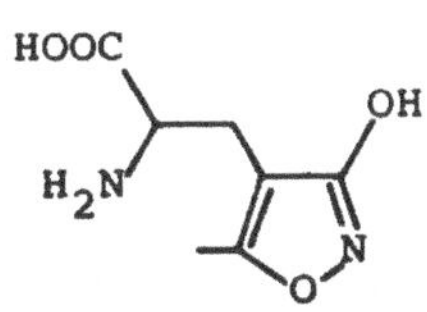	P	0	+	9, 14
Kainate	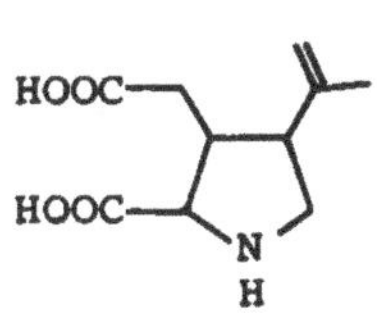	P	0	+	3, 5
Willaridine	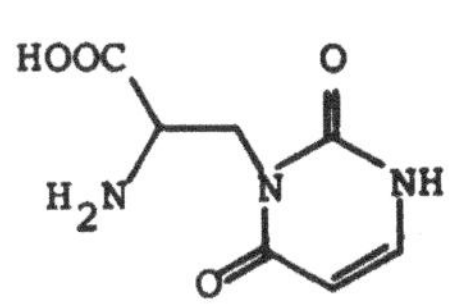	P	0	+	10
Quinolinate	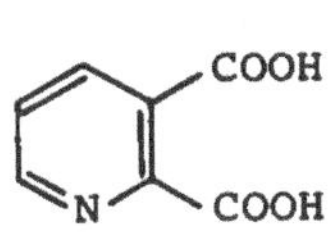	W	+	0	11
trans-PDA		W	+	0	12

Note: Symbols: 0 = inactive, + = active, W = weak, M = moderately potent, and P = potent.

Table 2
SELECTED EAA ANTAGONISTS

Compound[a]	Formula (nonionized form)	Antagonism of responses to NMDA	Quisqualate	Kainate	Ref.
D,L-APB	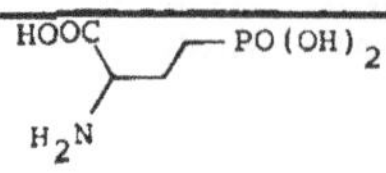	W	W	W	8, 22, 23, 28
D-APV	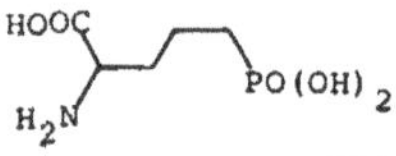	P	0	0	25—32
D-APH	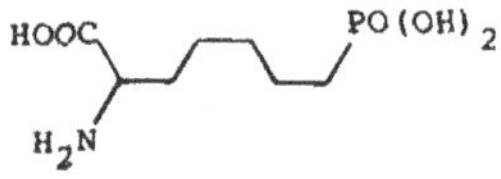	P	0	0	29
CPP	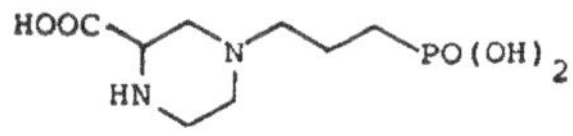	P	0	0	33
D-αAA	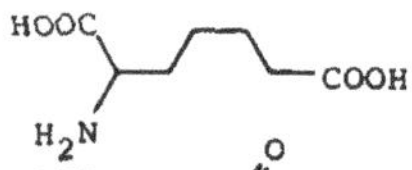	W	0	0	6, 7, 20, 21, 24
DGG	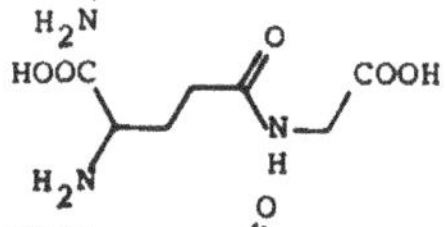	W	W	W	12
GAMS	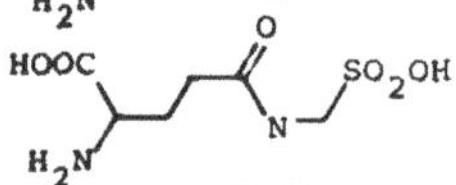	W	M	M	12, 32, 34
cis-PDA	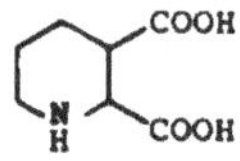	W	W	W	12, 43
cis-PzDA	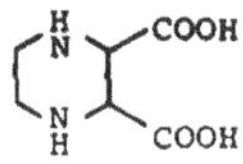	W	W	W	32, 44
pCB-PzDA	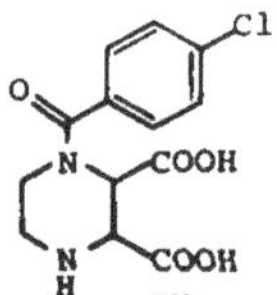	W	M	M	21, 44, 77
Kynurenate	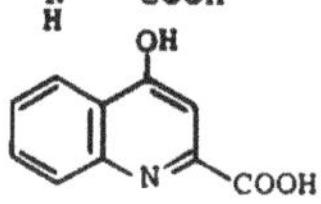	M	W	W	36—38
HQC	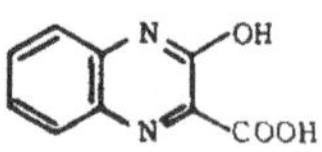	M	W	W	183
GDEE	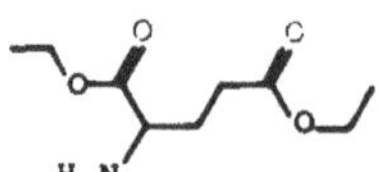	0	W	0	7—10, 18, 19, 22, 24
IIc	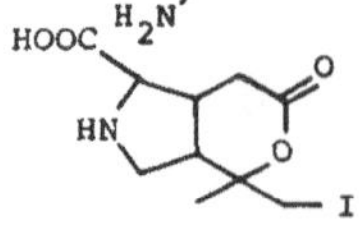	0	0	W	35
Ketamine[b]	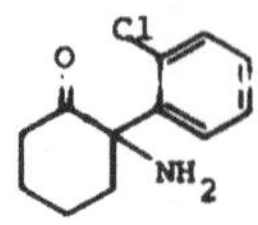	P	0	0	45, 47

Table 2 (continued)
SELECTED EAA ANTAGONISTS

Compound[a]	Formula (nonionized form)	Antagonism of responses to			Ref.
		NMDA	Quisqualate	Kainate	
MK 801[b]	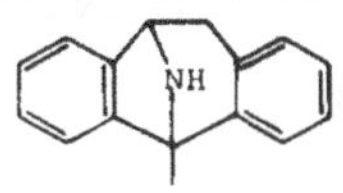	P	0	0	49, 50
PCP[b]	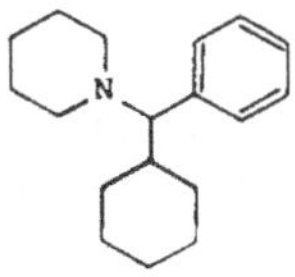	P	0	0	48

Note: Symbols: 0 = inactive, W ⟩ weak, M = moderately potent and P = potent.

[a] Abbreviations: PDA, 2,3-piperidine dicarboxylate; APB, 2-amino-4-phosphonobutyrate; APV, 2-amino-5-phosphonovalerate; APH, 2-amino-7-phosphonoheptanoate; CPP, 3-(2-carboxypiperazam-4-yl)-propyl-1-phosphonate; D-αAA, D-2-aminoadipate; DGG, γ-D-glutamylglycine; PzDA (±)-*cis*-2,3-piperazine dicarboxylate; pCB-PzDA, *N*-(*p*-chlorbenzoyl)piperazine-2,3-dicarboxylate; GAMS, γ-D-glutamylaminomethyl sulfonic acid; PCP, phencyclidine; HQC, 3-hydroxy-2-quinolinecarboxylate; GDEE, glutamatediethylester; IIc, kainate iodolacton; and MK 801, 5-methyl-10,11-dihydro-5H-dibenzo[a,d]cyclohepten-5,10-imine.

[b] Noncompetitive inhibition, probably via the NMDA ionophore.

sulfonic acid (GAMS) and certain lactonized kainate derivatives (e.g., Section II.C, Table 2) have shown selectivities against kainate responses.[32,34,35] Most other EAA antagonists including the endogenous substance kynurenic acid, certain dipeptides, such as γ-D-glutamylglycine (DGG), and piperidine or piperazine dicarboxylic acids, such as cis-PDA, cis-PzDA, and pCB-PzDA, are generally considered to be nonselective antagonists which block responses to both NMDA, quisqualate, and kainate.[32,36-44]

A distinct group of EAA antagonists, including ketamine, phencyclidine, and MK 801 belonging to the group of dissociative anaesthetics, have been shown to be potent and selective blockers of NMDA responses.[45-50] The substances do not displace ^{3}H-L-glutamate binding from the NMDA binding site, and the kinetics of their interaction with NMDA responses indicates a noncompetitive inhibition, probably via the NMDA-operated ionophore.[47-50]

B. Studies Using Intracellular Recordings

The ionic mechanisms by which EAAs exert their postsynaptic action have recently been studied carefully by several groups. Most of these studies have been performed in the rat of guinea pig hippocampal slice preparation with intracellular recordings from the pyrimidal cells in the CA_1 layer, but also patch clamp studies in neuronal cultures from the mouse striatum and spinal cord have given important contributions to our present knowledge of the ionophores coupled to EAA receptors.[51-68] From these studies it is evident that not only are NMDA receptors pharmacologically distinct from other EAA receptors, but also the ion channels operated by NMDA show characteristics different from the ionophores coupled to non-NMDA receptors.

Activation of non-NMDA receptors induces a fast depolarization of the postsynaptic neuronal membrane, accompanied by increases in membrane conductance.[51,54,56,57,62,66,67] These results were obtained under conditions where presynaptic effects were minimized and voltage-dependent sodium and calcium currents were largely suppressed (in the absence of Ca^{2+} and presence of Mn^{2+} and tetrodotoxin). The conductances due to activation of non-

Table 3
COMPARISON BETWEEN NMDA RESPONSES AND RESPONSES MEDIATED VIA NON-NMDA RECEPTORS

NMDA receptors	non-NMDA receptors
Generally not involved in monosynaptic responses, but involved in polysynaptic responses; induces "plastic" responses (e.g., LTP)	Induce fast response in practically all monosynaptic EAA pathways
Induce burst firing	No burst firing
Ionophore allows Na^+, K^+, and Ca^{2+} currents	Ionophore only permeable to Na^+ and K^+
Mg^{2+} gating of the ionophore	No Mg^{2+} sensitivity
Voltage dependent	No voltage dependency
Decreases or increases membrane conductance	Only increases in membrane conductance are seen
Maximal amplitude at resting potential	Amplitude increases with hyperpolarization
D-APV blocks competitively	D-APV inactive
Ketamine blocks noncompetitively	Ketamine inactive

NMDA receptors are voltage independent and have a reversal potential close to 0 mV.[66,67] These conductances involve Na^+ and K^+ currents (not Ca^{2+} currents) and seem to be quite similar to the currents induced by acetylcholine at the neuromuscular junction.[66,67]

The ionophore coupled to the NMDA receptor appears to have properties distinct from the non-NMDA receptor ionophores. The currents activated by NMDA seem to be carried primarily by Na^+ and K^+, but Ca^{2+} is most probably also permeable to the channel.[59,69] Moreover, the NMDA ionophore can be blocked by Mg^{2+} which has no effect on the non-NMDA receptor channels.[52,63-68] Patch-clamp analysis has revealed that in the absence of Mg^{2+}, NMDA channels are voltage independent.[63,64] However, when experiments are performed in the presence of Mg^{2+}, a characteristic voltage-dependent block of currents is seen.[63,64] The effect of Mg^{2+} is maximal at hyperpolarizing conditions and virtually absent at -20 mV and above.[63,64] At or below resting potential, the NMDA-activated channel is almost completely blocked by the physiological Mg^{2+} concentrations. However, when the membranes are depolarized, the Mg^{2+} blockade is "switched off". In Table 3, some of the distinct differences between NMDA responses and non-NMDA responses have been listed.

Investigations of a series of well-defined EAA pathways in the brain and spinal cord have revealed that monosynaptic responses (fast excitatory postsynaptic potentials) generally are generated via non-NMDA receptors.[70-79] Such responses cannot be blocked by the NMDA selective antagonist D-APV, but only by nonselective antagonists such as cis-PDA and GAMS (Table 2). However, NMDA has been shown to be involved in certain polysynaptic responses, and recent studies by Salt[78] and Herron et al.[68] indicate that NMDA may induce excitatory synaptic transmission dependent on the mode of stimulation.[68,72,74,78]

The Mg^{2+} block of NMDA receptors at resting potential may explain why synaptic NMDA responses may only be seen after a certain depolarization of the postsynaptic neurons, e.g., due to repetitive stimulation of non-NMDA receptors.[68,74] In agreement with this, NMDA has been shown to induce late EPSPs, e.g., in the hippocampal CA_1 pyramidal neurons.[74,79] NMDA also seems to be involved in the induction of the phenomenon long-term potentiation (LTP).[80-86] LTP is a sustained increase in synaptic efficacy following a brief tetanic stimulation of certain excitatory pathways.[80] LTP has been suggested to be involved in information storage, and it has recently been shown that NMDA antagonists such as D-APV which block LTP also may impair learning in animals.[84-87]

NMDA also induces epileptiform bursting activity in brain slices, and this phenomena as well as a similar effect seen after bathing of the slices in low Mg^{2+} concentrations can be blocked by D-APV.[53,88,89] In agreement with this, NMDA antagonists show antiepileptic activities in several models (Chapter 23).

C. Functional Neurochemical Studies

A series of neurochemical investigations on receptor-mediated effects of EAAs has added important information to the results from electrophysiological experiments. The induction of Na^+ currents by EAAs has been confirmed by studies of ^{22}Na fluxes in brain slices.[35,90] This model has been used to study pharmacological selectivities of EAA antagonists, and the results seem to correlate well with findings from electrophysiological studies.[35,90]

EAA-induced neuronal Ca^{2+} uptake has also been studied in brain slices and neuronal cultures.[91-95] Calcium influx may be through the NMDA-activated ionophore or through voltage-operated calcium channels opened when neurons are depolarized by any EAA.[59,69] It has recently been demonstrated that a calcium-activated kinase is concentrated in post-synaptic membranes of excitatory synapses.[96] The enzyme, which is not deactivated when the normal local concentration of calcium is reestablished, could be involved in an information storage process possibly related to LTP.[97] Although EAA-activated calcium uptake in this way may serve an important functional role, calcium over-load due to prolonged EAA activation may be deleterious to the cell. Calcium overload leads to severe disturbances in cellular metabolism and to activation of proteases and lipases in the cell.[98]

The EAA-induced local accumulation of Ca^{2+} may function as a secondary messenger system. However, EAAs have also been shown to activate cGMP formation and polyphosphoinositol turnover in brain slices and neuronal cultures.[42,99-103] Although these effects may be unspecific and simply secondary responses to depolarizations of the neurons, they may on the other hand represent triggers for cascade signals involved in long-term modulation of synaptic function.

As a consequence of neuronal depolarization, EAAs are able to induce neurotransmitter release. This has been demonstrated for a number of neurotransmitters both in vivo, e.g., using the push-pull cannula technique, and in vitro in brain slices and neuronal cultures. In vitro models of acetylcholine release from brain slices and GABA release from cultured neurons have been used for pharmacological studies of EAA antagonist selectivities.[104,105]

III. RECEPTOR BINDING STUDIES ON MEMBRANE PREPARATIONS

A. ^{3}H-L-Glutamate Binding

The first glutamate binding studies were published independently by Roberts and by Michaelis et al.[106,107] using ^{14}C-glutamate as the radioligand. ^{14}C-glutamate binding was found to be saturable, with a relatively high affinity, and displaceable by other EAAs.[106,107] When ^3H-L-glutamate with a much higher specific activity became available, more detailed glutamate binding studies could be performed and a number of groups started developing ^3H-L-glutamate binding assays.[108-126] Due to the fact that most of these studies started almost at the same time, a number of quite different approaches were used, including different membrane preparations, with or without freezing/thawing steps and with a number of different buffers. These studies have been carefully reviewed by Foster and Fagg.[127] Although the quite different results obtained using different methodologies seemed very confusing a few years ago, it is now possible to see trends in the results. The dependency of ^3H-L-glutamate binding on the membrane preparations and buffers is summarized below and shown in a schematic form in Table 4.

^3H-L-glutamate binding to fresh homogenates or homogenates made from whole frozen tissue is maximal when experiments are performed in the presence of Na^+.[127] Under these conditions, the major amount of ^3H-L-glutamate binds to a site which shows pharmacological

Table 4
^{3}H-L-GLUTAMATE BINDING ASSAYS ON DIFFERENT MEMBRANE PREPARATIONS

	Na$^+$ independent		
Membrane preparation	**Cl$^-$/Ca^{2+} independent**	**Cl$^-$/Ca^{2+} dependent**	**Na$^+$ dependent**
Crude homogenate and synaptosomal membrane preparation[a]			
Fresh	$A_1 + A_2$	A_4	U
Frozen (whole brain)	$A_1 + A_2$	A_4[b]	U
Frozen (homogenate)	$A_1 + A_2$	0	U[c]
PSD fraction			
Fresh	$A_1 + A_2$	0	0
Frozen	$A_1 + A_2$	0	0

Note: A_1, NMDA binding site; A_2, quisqualate binding site; A_4, probably uptake into presynaptic vesicles (L-APB sensitive site); and U, binding site for glutamate carrier.

[a] Five or more wash steps including incubation at 37°C.
[b] Freezing eliminates about 30% of binding.
[c] Freezing eliminates about 50% of binding.

selectivities similar to those of the high-affinity glutamate uptake carrier in neuronal membranes.[128,129]

When ^{3}H-L-glutamate binding is studied in the absence of Na$^+$, binding to the uptake carrier is abolished and ^{3}H-L-glutamate binds almost exclusively to the so-called A_4 binding site.[127] This site is characterized by a high sensitivity to the ligand L-APB and binding is increased significantly in the presence of Cl$^-$ and Ca^{2+}.[117]

Freezing of homogenates will remove 80 to 90% of the ^{3}H-L-glutamate binding.[122] The remaining binding sites are independent of Cl$^-$ and the high sensitivity to L-2-amino-4-phosphonobutyrate (L-APB) is lost.[122,126] However, the remaining binding sites seem to correlate with NMDA (A_1) and quisqualate (A_2) receptors (around 50% of each).[126] Recently it has been shown by Fagg and Matus[125] that the same binding sites are revealed when ^{3}H-L-glutamate binding is performed on a preparation of so-called postsynaptic densities (PSD fraction).[125,130] In this preparation, neither the A_4 binding site nor the Na$^+$-dependent binding to the transport carrier is seen.[125] This indicates that both the A_4 binding site and the glutamate carrier are localized in presynaptic or extrasynaptic elements.

In the preparations where ^{3}H-L-glutamate binds only to A_1 and A_2 sites (PSD fractions or frozen crude homogenates), it is possible to study ^{3}H-L-glutamate binding to the NMDA receptor (A_1) alone simply by blocking the quisqualate receptor with low concentrations (10 μM) of quisqualate.[125] Similarily, quisqualate receptors (A_2) can be studied in such preparations by blocking NMDA receptors with 100 μ*M* NMDA. However, some overlap in the selectivities of NMDA and quisqualate reduces the apparent affinities of the binding site for the tested substances.[125]

The possible role of the Cl$^-$/Ca^{2+}-dependent ^{3}H-L-glutamate binding sites (A_4) is still unresolved. Although L-APB is a potent displacer of ^{3}H-L-glutamate from this site, the binding site does not correlate with the L-APB sensitive synaptic responses to glutamate in the hippocampal slice.[131] Recent studies by Pin et al.[132] strongly suggest that the A_4 site corresponds to an uptake of ^{3}H-L-glutamate into presynaptic vesicles .[132] This may also explain why the A_4 binding sites are missing in frozen prepations.[122] Uptake into presynaptic vesicles is also in agreement with results of a recent study of ^{3}H-D-aspartate binding to rat brain membranes.[133] Furthermore, the inactivation curve of ^{3}H-L-glutamate binding after high-energy irradiation is in favor of this hypothesis.[134]

A summary of the results of ^{3}H-L-glutamate binding studies using a series of different assay conditions is given in Table 4. Using selective assay conditions, ^{3}H-L-glutamate binding may be used to study binding sites which seem to correlate to the NMDA receptors and quisqualate receptors.[125,126] However, kainate receptors cannot be studied using ^{3}H-L-glutamate as the radioligand since kainate for some unknown reason has never been shown to displace ^{3}H-L-glutamate binding regardless of the membrane preparation used.[127]

B. Selective Radioligands

1. NMDA Receptors (A_1 Sites)

Although the NMDA receptor has been characterized extensively in neuropharmacological experiments, it has proven difficult to develop a high-affinity NMDA selective radioligand for receptor binding studies. As mentioned above, NMDA receptors can be studied using the radioligand ^{3}H-L-glutamate (in the presence of 10 μM quisqualate).[125] This is, of course, a rather indirect measure of the A_1 binding sites, and several attempts at developing a more selective A_1 radioligand have been made. NMDA itself has been radiolabeled, but in binding experiments ^{3}H-NMDA proved to have very low affinity.[135] Studies using the specific NMDA antagonists ^{3}H-D-APV and ^{3}H-D-APH have encountered the same experimental problems, although these ligands seem to be superior to NMDA.[136,137] As described below, ^{3}H-D-APV has successfully been used in autoradiographic studies to label NMDA receptors.[138]

Most recently Watkins and colleagues[33] have developed a cyclic analogue of D-APH called CPP. This substance which also is a highly selective NMDA antagonist has been radiolabeled and preliminary studies by Watkins and co-workers[184] in our laboratory and by Murphy et al.[139] indicate that ^{3}H-CPP binds with high affinity to NMDA receptors in brain membrane preparations with low nonspecific binding. As previously mentioned, dissociative anaesthetics such as ketamin and phencyclidin (PCP) act as noncompetitive antagonists of NMDA responses, probably by interaction with the NMDA-operated ionophore.[47,50] Radiolabeled compounds with this mechanism of action may prove to be valuable tools in biochemical studies of allosteric interactions between the NMDA receptor and its ionophore. Similar binding studies using compounds directed toward the Cl^- channels (e.g., ^{35}S-t-butylbicyclophosphothionate, ^{35}S-TBPS) have given valuable information on the allosteric interactions in the GABA-benzodiazepine-Cl^- channel receptor complex. Binding studies with ^{3}H-PCP on membrane preparations have been limited by the high binding of the compound to glassware, etc. The PCP analogue ^{3}H-TCP, *N*-(1-[2-thienyl] cyclohexyl-3,4-piperidine), which was developed as a sigma opiate receptor ligand has recently also been used as a ligand for NMDA ionophores in autoradiographic binding studies.[140] ^{3}H-TCP also seems to be useful for binding studies on the NMDA-ionophore in membrane preparations.[141] An alternative radioligand for binding studies on the NMDA-ionophore is the substance ^{3}H-MK 801 (Table 2) which is claimed to be more selective for the NMDA ionophore.[50]

2. Quisqualate Receptors (A_2 Sites)

The synthetic EAA AMPA (Table 1) developed by Krogsgaard-Larsen and co-workers[9] has proved to be a very selective agonist at the quisqualate receptor with a potency comparable to quisqualate. On the basis of this, ^{3}H-AMPA was an obvious choice as a radioligand for studies of quisqualate receptors. ^{3}H-AMPA binds with high affinity to brain membrane preparations.[142-145] The nonspecific binding is relatively low, but the number of ^{3}H-AMPA binding sites is also low.[142] However, it has been found that addition of the chaotropic ion SCN^- leads to a dramatic increase in affinity of ^{3}H-AMPA for the binding sites.[144] It has been shown that SCN^- ions change the equilibrium of a two-state quisqualate receptor

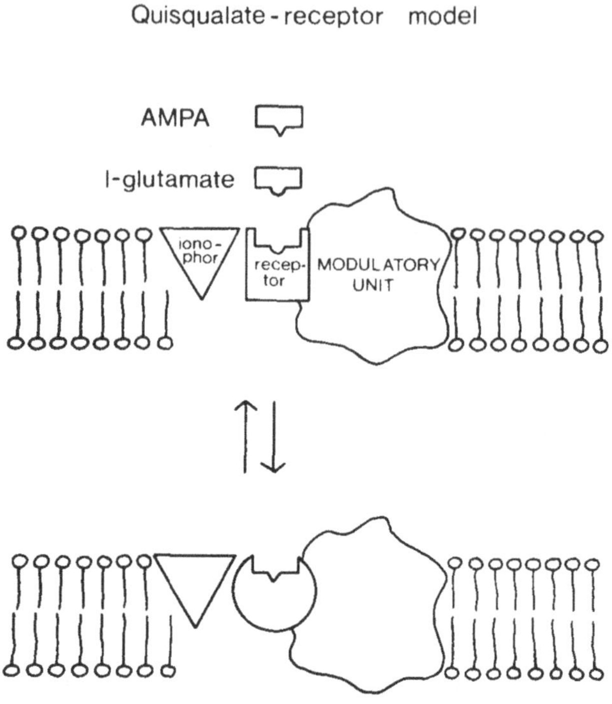

FIGURE 1. Hypothetical model for quisqualate receptors. The hypothesis is based on results in References 144 and 145. Quisqualate receptors exist in two states, one with high affinity for AMPA and one with low affinity for AMPA, but with high affinity for L-glutamate. The equilibrium between the two states can be affected by SCN^- ions. Some compounds, e.g., quisqualate, have high affinity for both states. A modulatory site is coupled to both states. Removal of the modulatory site, e.g., by high-energy irradiation, enhances the affinity of both sites.

against the conformation with high affinity for ^{3}H-AMPA.[145] Radiation inactivation analyses on the ^{3}H-AMPA binding site has shown that the binding site consists of two macromolecules of different molecular weights.[144] Removal of the high molecular weight molecule by irradiation increases the specific binding of ^{3}H-AMPA.[144] A hypothetical model of the quisqualate receptor is suggested in Figure 1.

As mentioned earlier, quisqualate receptors can also be studied using ^{3}H-L-glutamate as the radioligand in the presence of NMDA.[125] However, binding assays using the specific ligand ^{3}H-AMPA in the presence of SCN^- appear to be a more direct and reliable model for studies of quisqualate receptor binding sites.[144]

3. Kainate Receptors

The only radioligand used for binding studies on the kainate receptor is ^{3}H-kainate. ^{3}H-kainate binding is insensitive to freezing and relatively independent of Na^+.[146-151] Scatchard analysis of ^{3}H-kainate binding performed in the absence of Ca^{2+} revealed a high- and a low-affinity binding site.[148,150] High-affinity ^{3}H-kainate binding sites may correlate to kainate receptors studied in electrophysiological experiments.[127]

In the presence of Ca^{2+} only the low-affinity binding site is seen, and it has been suggested that these ^{3}H-kainate binding sites may correspond to quisqualate receptors.[152] This assumption is based on the pharmacological specificities of the high- and low-affinity ^{3}H-kainate binding sites and on recent results from radiation inactivation analyses of the ^{3}H-

Table 5
EAA RECEPTOR BINDING TO BRAIN MEMBRANE PREPARATIONS

		Non-NMDA receptors	
	NMDA receptors (A_1)	**Quisqualate receptors (A_2)**	**Kainate receptors (A_3)**
Selective radioligands	^{3}H-L-glutamate (+10 μM QUIS)	^{3}H-AMPA	^{3}H-kainate
	^{3}H-D-APV	^{3}H-L-glutamate (+100 μM NMDA)	
	^{3}H-CPP		
	(^{3}H-MK 801)		
	(^{3}H-TCP)		
Potent displacers[a]	L-Glutamate	Quisqualate	Kainate
	L-Aspartate	AMPA	Quisqualate
	D-APV	L-Glutamate	L-Glutamate
	Ibotenate	Kainate	Ibotenate
	NMDA		
Weak or inactive displacers	Quisqualate	Ibotenate	AMPA
	AMPA	L-Aspartate	L-Aspartate
	Kainate	NMDA	NMDA

[a] The inhibitors have been listed according to potencies using the following radioligands: NMDA receptors, ^{3}H-L-glutamate (in the presence of 10 μM quisqualate); quisqualate receptors, ^{3}H-AMPA (in the presence of 100 mM KSCN); and kainate receptors, ^{3}H-kainate (in the absence of calcium ions).

kainate binding sites.[152] It was found that the molecular weight of the low-affinity ^{3}H-kainate binding site (studied in the presence of Ca^{2+}) was similar to that of the ^{3}H-AMPA site.[142] As in the case of the ^{3}H-AMPA binding site, the low-affinity ^{3}H-kainate binding could be increased by denaturing (irradiation) the high molecular weight molecule from the complex.[142] Furthermore, autoradiographic binding studies (see below) indicate that quisqualate sites may be distinguished from kainate sites by their sensitivity to calcium ions. Similarities and differences between quisqualate and kainate receptors are given in Table 5.

It should be noted that quisqualate in contrast to AMPA is a relatively potent displacer of ^{3}H-kainate binding.[142] Since AMPA and not quisqualate is the most selective ligand for the quisqualate receptor, it has been suggested that the receptor should be renamed (e.g., A_2, AA_2, or AMPA receptor).

C. Other Ligands

A series of other more or less selective radioligands has been used for studies of EAA binding sites:

1. ^{3}H-L-aspartate labels a binding site which shows some similarities to the Cl^-/Ca^{2+} independent ^{3}H-L-glutamate binding site. However, only a few reports on the structural requirements of the binding site have appeared.[153-156]
2. ^{3}H-DL-APB binds to a distinct binding site probably equivalent to the Cl^-/Ca^{2+}-dependent APB-sensitive ^{3}H-L-glutamate binding site. The binding site may represent uptake into presynaptic vesicles.[132,157,158]

Table 6
NON-NMDA RECEPTORS. SIMILARITIES AND DIFFERENCES BETWEEN QUISQUALATE AND KAINATE RECEPTORS

Similarities	Differences
Quisqualate is an inhibitor of ^{3}H-AMPA binding as well as ^{3}H-kainate binding to rat cortex membranes	^{3}H-AMPA binding is increased by chaotropic ions, whereas ^{3}H-kainate binding is unaltered
Kainate is an inhibitor of ^{3}H-kainate binding as well as ^{3}H-AMPA binding to rat cortex membranes	Calcium ions increase ^{3}H-AMPA binding, whereas ^{3}H-kainate binding is dramatically decreased
GAMS reduces responses induced by quisqualate as well as kainate in the cat spinal cord	AMPA is inactive as an inhibitor of kainate binding
Seizures in DBA/2-mice after icv administration of kainate or quisqualate are equally well blocked by GAMS	Autoradiographic studies have shown different regional distribution of ^{3}H-AMPA sites and ^{3}H-kainate sites
Radiation inactivation curves for ^{3}H-kainate binding in the presence of calcium ions are similar to ^{3}H-AMPA inactivation curves	Radiation inactivation analysis of the ^{3}H-AMPA binding site and the high-affinity ^{3}H-kainate binding site gives different molecular target sizes

3. L-^{35}S-cysteate and L-^{3}H-cysteine sulfinate may bind to identical binding sites, but the relationship to binding sites labeled by other EAA is at present unclear.[159,160]

Table 6 gives a summary of radioligands used to characterize NMDA, quisqualate, and kainate receptors with indications of pharmacological selectivities of selected ligands.

IV. AUTORADIOGRAPHIC BINDING STUDIES

A few years ago the autoradiographic receptor binding technique was introduced for studies of the different EAA receptor subtypes in the brain. This technique offered the possibility of studying the regional distribution of each receptor subtype with a resolution at the cellular level. Such information could ultimately provide the important missing link between receptor binding studies in brain homogenates and findings in vivo in electrophysiological studies as well as in neuropathological studies.

The first reports of ^{3}H-L-glutamate binding using quantitative autoradiography were published in 1983 by Greenamyre et al.,[161] Halpain et al.,[162] and Monaghan et al.[163] All groups described a high-affinity, saturable, and reversible binding of ^{3}H-L-glutamate in brain slices.[161-163] Monaghan et al.[163] restricted their studies to the hippocampus where they found four different ^{3}H-L-glutamate binding sites depending on the assay conditions.[163] In the absence of $CaCl_2$, most of the ^{3}H-L-glutamate could be displaced by NMDA.[163] The remaining bound ^{3}H-L-glutamate could in some regions be displaced selectively by kainate (e.g., in the stratum lucidum of the hippocampal CA_3 layer), whereas in other regions quisqualate and AMPA were selective displacers of ^{3}H-L-glutamate binding, (e.g., in the pyramidal CA_1 layer).[163] This indicates that in the hippocampal slice ^{3}H-L-glutamate binds to all three receptor types: the NMDA, quisqualate, and kainate receptors. It is interesting that kainate in these autoradiographic binding studies in contrast to binding studies on homogenates was able to displace ^{3}H-L-glutamate binding. The kainate-sensitive ^{3}H-L-glutamate binding in the hippocampal slices could be decreased by Ca^{2+} in agreement with the results from ^{3}H-kainate binding studies in membrane preparations.[163] In addition to the three classical binding sites, a fourth atypical ^{3}H-L-glutamate binding site was revealed in the hippocampal slice by the addition of $CaCl_2$.[163] This binding site is discussed in Section V.

The original finding in the hippocampal slice that ^{3}H-L-glutamate binding in the absence of $CaCl_2$ largely (approximately 80%) represents NMDA receptors has recently been confirmed in other brain regions.[164-168] In agreement with this, ^{3}H-D-APV which selectively labels NMDA receptors shows the same regional distribution as ^{3}H-L-glutamate binding (studied without $CaCl_2$).[138] Furthermore, it has been shown by autoradiography that the regional binding of ^{3}H-TCP, which is assumed to label the NMDA-ionophore, correlates with NMDA binding sites.[168]

As mentioned earlier, ^{3}H-L-glutamate may also be used to study the regional distribution of quisqualate receptors.[163,166] These binding sites are revealed when high concentrations of NMDA and Ca^{2+} are added to block NMDA- and high-affinity kainate receptors.[163,166] A more direct measure of the regional distribution of quisqualate receptors can be obtained using the highly selective radioligand ^{3}H-AMPA. Autoradiographic studies of ^{3}H-AMPA binding have shown a good correlation with the localization of quisqualate binding sites estimated from experiments using ^{3}H-L-glutamate as the radioligand.[169-171]

Autoradiographic studies of ^{3}H-kainate binding reveal a pattern of grains similar to the distribution of kainate-sensitive ^{3}H-L-glutamate binding.[163,166,172-174] A high density of ^{3}H-kainate binding, e.g., in the stratum lucidum CA_3 pyramidal cells of the hippocampus, is in agreement with the selective vulnerability for injected kainate in this cell layer.[173]

V. ATYPICAL RECEPTORS

Not all experimental findings on EAA receptors from neurochemical and neuropharmacological studies fit into the general classification scheme discussed previously. In the mudpuppy retina, L-APB is a potent agonist on ON-bipolar cells, mimicking the effects of L-glutamate.[175,176] The responses are sensitive to blockade by the NMDA antagonist D-APV, although NMDA is inactive in this system.[176] L-APB has in other models shown potent antagonism of synaptic responses (probably EAA mediated).[28,29,177,178] However, L-APB at the same concentrations showed no blockade of responses to iontophoretically supplied L-glutamate, NMDA, quisqualate, or kainate.[28,29,177,178] It has recently been shown that brief exposure of hippocampal slices to quisqualate increases the potency of L-APB for the depression of stimulus evoked extracellular synaptic field potentials.[179] Altogether, L-APB interaction with EAA neurotransmission appears to be complex and far from resolved today.

As mentioned above, an atypical ^{3}H-L-glutamate binding site has been found in autoradiographic studies in the hippocampal slice.[163] Binding to this site which did not appear to be an uptake site could only be displaced by L-glutamate and not by, for example, L-APB.[163] This distinct binding site may correspond to a so-called Glu B binding site, recently described by Nadler and co-workers[180,181] in a hippocampal membrane preparation, and to a ^{3}H-L-glutamate binding site on cerebellar granule cells described by Drejer et al.[182] In the granule cells, the receptor appeared to be functionally active since only L-glutamate and D,L-aspartate among a series of agonists tested were able to induce a D-APV-sensitive ^{3}H-D-aspartate release from the neurons.[182]

VI. CONCLUDING REMARKS

During the last few years, EAAs have been identified as the transmitters in a large number of excitatory pathways in the brain and spinal cord. Fast monosynaptic responses are mediated largely or solely via the non-NMDA receptors, but NMDA also appears to be involved in some synaptic responses as well as in regulatory functions such as LTP.

Recent electrophysiological studies of the ionic mechanisms of EAA responses in nerve cells, using intracellular recordings and patch-clamp analysis, have given detailed information on the mechanisms of EAA action. Perhaps most importantly, it has been shown that the NMDA receptor exhibits a characteristic Mg^{2+}-gated voltage dependency. These findings indicate that EAA responses are not merely on/off processes, but involve state-dependent long-term effects subject to local regulation.

Of the series of newly developed EAA antagonists, it has only been possible to identify potent and selective antagonists for the NMDA receptor subtype. Responses to quisqualate and kainate may be blocked by other antagonists, but all of these are weak and rather unselective.

Specific receptor binding assays for the different EAA receptor subtypes are important tools as screening models for the development of new antagonists. Quisqualate and kainate receptors can be studied simply and reliably in a frozen crude membrane preparation using the radioligands ^{3}H-AMPA and ^{3}H-kainate, respectively. In the same membrane preparation, NMDA receptors can be studied indirectly using ^{3}H-L-glutamate as the radioligand. The new ligand ^{3}H-CPP may solve the remaining problem of developing a selective high-affinity radioligand for binding studies on the NMDA receptor.

Recent autoradiographic binding studies have given detailed information on the regional distribution and ontogenetic development of the different subtypes of EAA receptors in the CNS. The findings from these studies show a good correlation with results from membrane binding studies and electrophysiological studies and substantiate the original classification scheme for EAA receptors.

As discussed in Chapters 23, a number of neurological diseases including pathological nerve cell degenerations most probably involve an uncontrolled hyperactivity in EAA pathways. A number of selective and nonselective NMDA antagonists have shown prominent antiepileptic effects and effects against glutamate-, hyperglycemia-, or ischemia-induced nerve cell degenerations in different models. However, all antagonists published so far are highly polar and pass the blood-brain barrier only poorly. A major goal for the years to come is to develop potent and selective EAA antagonists with chemical properties that allow transport into the brain. Such compounds may be of considerable value not only as pharmacological tools for further characterizations of EAA responses, but also as potential therapeutics for treatments of diseases involving EAA hyperactivity.

REFERENCES

1. **Curtis, D. R., Phillis, J. W., and Watkins, J. C.,** The chemical excitation of spinal neurons, *Nature (London),* 183, 611, 1959.
2. **Curtis, D. R. and Watkins, J. C.,** Acidic amino acids with strong excitatory actions on mammalian neurons, *J. Physiol. (London),* 166, 1, 1963.
3. **Shinozaki, H. and Konishi, S.,** Actions of several anthelmintics and insecticides on rat cortical neurones, *Brain Res.,* 24, 368, 1970.
4. **Johnston, G. A. R., Curtis, D. R., Davies, J., and McCulloch, R. M.,** Spinal interneurone excitation by conformationally restricted analogues of L-glutamate, *Nature (London),* 248, 804, 1974.
5. **Shinozaki, H. and Shibuya, I.,** A new potent excitant quisqualic acid; effects on crayfish neuromuscular junction, *Neuropharmacology,* 13, 665, 1974.
6. **Biscoe, T. J., Evans, R. H., Headley, P. M., Martin, M. R., and Watkins, J. C.,** Structure-activity relations of excitatory amino acids on frog and rat spinal neurones, *Br. J. Pharmacol.,* 58, 373, 1976.
7. **Hicks, T. P., Hall, J. G., and McLennan, H.,** Ranking of excitatory amino acids by the antagonists glutamic acid diethyl ester and D-α-aminoadipic acid, *Can. J. Physiol. Pharmacol.,* 56, 901, 1978.
8. **Davies, J., Evans, R. H., Francis, A. A., and Watkins, J. C.,** Excitatory amino acid receptors and synaptic excitation in the mammalian central nervous system, *J. Physiol. (Paris),* 75, 641, 1979.

9. **Krogsgaard-Larsen, P., Honoré, T., Hansen, J. J., Curtis, D. R., and Lodge, D.,** New class of glutamate agonist structurally related to ibotenic acid, *Nature (London)*, 284, 64, 1980.
10. **Evans, R. H., Jones, W., and Watkins, J. C.,** Willaridine: a potent quisqualate-like excitant, *J. Physiol. (London)*, 308, 71P, 1981.
11. **Stone, T. W. and Perkins, M. N.,** Quinolinic acid: a potent endogenous excitant at amino acid receptors in rat CNS, *Eur. J. Pharmacol.*, 72, 411, 1981.
12. **Davies, J., Evans, R. H., Jones, A. W., Smith, D. A. S., and Watkins, J. C.,** Differential activation and blockade of excitatory amino acid receptors in the mammalian and amphibian central nervous systems, *Comp. Biochem. Physiol.*, 72, 211, 1982.
13. **Takeuchi, H., Watanabe, K., Nomoto, K., Ohfune, Y., and Takemoto, T.,** Effects of α-kainic acid, domoic acid and their derivatives on a molluscan giant neuron sensitive to β-hydroxy-L-glutamic acid, *Eur. J. Pharmacol.*, 102, 325, 1984.
14. **Krogsgaard-Larsen, P., Nielsen, E. Ø., and Curtis, D. R.,** Ibotenic acid analogues. Synthesis and biological and in vitro activity of conformationally restricted agonists at central excitatory amino acid receptors, *J. Med. Chem.*, 27, 585, 1984.
15. **Agrawal, S. G. and Evans, R. H.,** The primary afferent depolarizing action of kainate in the rat, *Br. J. Pharmacol.*, 87, 345, 1986.
16. **Watkins, J. C. and Evans, R. H.,** Excitatory amino acid transmitters, *Annu. Rev. Pharmacol. Toxicol.*, 21, 165, 1981.
17. **McLennan, H.,** Receptors for the excitatory amino acids in the mammalian central nervous system, *Prog. Neurobiol.*, 20, 251, 1983.
18. **Haldeman, S. and McLennan, H.,** The antagonistic action of glutamate diethyl ester towards amino acid-induced and synaptic excitations of central neurons, *Brain Res.*, 45, 393, 1972.
19. **Spencer, H. J.,** Antagonism of cortical excitation of striatal neurons by glutamic acid diethyl ester: evidence for glutamate as an excitatory transmitter in the rat striatum, *Brain Res.*, 102, 91, 1976.
20. **Biscoe, T. J., Davies, J., Dray, A., Evans, R. H., Francis, A. A., Martin, M. R., and Watkins, J. C.,** Depression of synaptic excitation and of amino acid-induced excitatory responses of spinal neurones by d-α-aminoadipate, α,ε-diaminopimelic acid and HA 966, *Eur. J. Pharmacol.*, 45, 315, 1977.
21. **Evans, R. H., Francis, A. A., and Watkins, J. C.,** Mg^{2+}-like selective antagonism of excitatory amino acid-induced responses by α,ε-diaminopimelic acid, D-α-aminoadipate and HA-966 in isolated spinal cord of frog and immature rat, *Brain Res.*, 148, 536, 1978.
22. **Davies, J. and Watkins, J. C.,** Selective antagonism of amino acid-induced and synaptic excitation in the cat spinal cord, *J. Physiol.*, 297, 621, 1979.
23. **Evans, R. H., Francis, A. A., Hunt, K., Oakes, D. J., and Watkins, J. C.,** Antagonism of excitatory amino acid-induced responses and of synaptic excitation in the spinal cord of the frog, *Br. J. Pharmacol.*, 67, 591, 1979.
24. **McLennan, H. and Lodge, D.,** The antagonism of amino acid-induced excitation of spinal neurones in the cat, *Brain Res.*, 1969, 83, 1979.
25. **Davies, J., Francis, A. A., Jones, A. W., and Watkins, J. C.,** 2-Amino-5-phosphonovalerate (2APV), a potent and selective antagonist of amino acid-induced and synaptic excitation, *Neurosci. Lett.*, 21, 77, 1981.
26. **Perkins, M. N., Stone, T. W., Collins, J. F., and Curry, K.,** Phosphonate analogues of carboxylic acids as amino acid antagonists on rat cortical neurons, *Neurosci. Lett.*, 23, 333, 1981.
27. **Stone, T. W., Perkins, M. N., Collins, J. F., and Curry, K.,** Activity of the enantiomers of 2-amino-5-phosphonovaleric acid as stereospecific antagonists of excitatory amino acids, *Neuroscience*, 6, 2249, 1981.
28. **Davies, J. and Watkins, J. C.,** Actions of the D- and L-forms of 2-amino-5-phosphonovalerate and 2-amino-4-phosphonobutyrate in the cat spinal cord, *Brain Res.*, 235, 378, 1982.
29. **Evans, R. H., Francis, A. A., Jones, A. W., Smith, D. A. S., and Watkins, J. C.,** The effect of a series of ω-phosphonic-α-carboxylic amino acids on electrically evoked and excitant amino acid-induced responses in isolated spinal cord preparations, *Br. J. Pharmacol.*, 75, 65, 1982.
30. **McLennan, H. and Liu, J.,** The action of six antagonists of the excitatory amino acids on neurons of the rat spinal cord, *Exp. Brain Res.*, 45, 151, 1982.
31. **Corradetti, R., King, A. E., Nistri, A., Rovira, C., and Sivilotti, L.,** Pharmacological characterization of D-aminophosphonovaleric acid antagonism of amino acid and synaptically evoked excitations on frog motoneurones in vitro: an intracellular study, *Br. J. Pharmacol.*, 86, 19, 1985.
32. **Davies, J., Jones, A. W., Sheardown, M. J., Smith, D. A. S., and Watkins, J. C.,** Phosphono dipeptides and piperazine derivatives as antagonists of amino acid-induced and synaptic excitation in mammalian and amphibian spinal cord, *Neurosci. Lett.*, 52, 79, 1984.
33. **Davies, J., Evans, R. H., Herrling, P. L., Jones, A. W., Olverman, H. J., Pook, P., and Watkins, J. C.,** CPP, a new potent and selective NMDA antagonist. Depression of central neuron responses, affinity for (^{3}H) d-AP5 binding sites on brain membranes and anticonvulsant activity, *Brain Res.*, 382, 169, 1986.

34. **Davies, J. and Watkins, J. C.,** Depressant actions of γ-D-glutamylaminomethylsulphonate (GAMS) on amino acid induced and synaptic excitation in the cat spinal cord, *Brain Res.*, 327, 113, 1985.
35. **Goldberg, O., Luini, A., and Teichberg, V. I.,** Lactones from kainic acid: novel selective antagonists of amino acid-induced Na^+ fluxes in rat striatum slices, *Neurosci. Lett.*, 23, 187, 1981.
36. **Ganong, A. H., Lanthorn, T. H., and Cotman, C. W.,** Kynurenic acid inhibits synaptic and acidic amino acid-induced responses in the rat hippocampus and spinal cord, *Brain Res.*, 273, 170, 1983.
37. **Perkins, M. N. and Stone, T. W.,** A study of kunurenic acid and excitatory amino acids in the rat hippocampus, *J. Physiol. (London)*, 357, 118P, 1984.
38. **Robinson, M. K., Anderson, K. D., and Koerner, J. F.,** Kynurenic acid as an antagonist of hippocampal excitatory transmission, *Brain Res.*, 309, 119, 1984.
39. **Francis, A. A., Jones, A. W., and Watkins, J. C.,** Dipeptide antagonists of amino acid-induced and synaptic excitation in the frog spinal cord, *J. Neurochem.*, 35, 1458, 1980.
40. **Davies, J. and Watkins, J. C.,** Differentiation of kainate and quisqualate receptors in the spinal cord by selective antagonist with γ-D (and L) -glutamylglycine, *Brain Res.*, 206, 172, 1981.
41. **Sawada, S. and Yamamoto, C.,** Gamma-D-glutamylglycine and cis-2,3-piperidine dicarboxylate as antagonists of excitatory amino acids in the hippocampus, *Exp. Brain Res.*, 55, 351, 1984.
42. **Foster, G. A. and Roberts, P. J.,** Pharmacology of excitatory amino acid receptors mediating the stimulation of rat cerebellar cyclic GMP levels in vitro, *Life Sci.*, 27, 215, 1980.
43. **Davies, J., Evans, R. H., Francis, A. A., Jones, A. W., and Watkins, J. C.,** Antagonism of excitatory amino acid-induced and synaptic excitation of spinal neurons by cis-2,3-piperidine dicarboxylate, *J. Neurochem.*, 36, 1305, 1981.
44. **Ganong, A. H., Jones, A. W., Watkins, J. C., and Cotman, C. W.,** Parallel antagonism of synaptic transmission and kainate/quisqualate responses in the hippocampus by piperazine-2,3-dicarboxylic acid analogs, *J. Neurosci.*, 6, 930, 1986.
45. **Anis, N. A., Berry, S. C., Burton, N. R., and Lodge, D.,** The dissociative anaesthetics, ketamine and phencyclidine, selectively reduce excitation of central mammalian neurones by N-methyl-aspartate, *Br. J. Pharmacol.*, 79, 565, 1983.
46. **Berry, S. C. and Lodge, D.,** Benz(f)isoquinolines as excitatory amino acid antagonists: an indication of their method of action?, *Biochem. Pharmacol.*, 33, 3829, 1984.
47. **Thompson, A. M., West, D. C., and Lodge, D.,** An N-methylaspartate receptor-mediated synapse in rat cerebral cortex: a site of action of ketamine?, *Nature (London)*, 313, 479, 1985.
48. **Aanonsen, L. M. and Wilcox, G. L.,** Phencyclidine selectively blocks a spinal action of N-methyl-D-aspartate in mice, *Neurosci. Lett.*, 67, 191, 1986.
49. **Kemp, J. A., Priestley, T., and Woodruff, G. N.,** MK-801, a novel, orally active anticonvulsant, is a potent, non-competitive N-methyl-D-aspartate-receptor antagonist, *Br. J. Pharmacol.*, 89, 535P, 1986.
50. **Wong, E. H. F., Kemp, J. A., Priestley, T., Knight, A. R., Woodruff, G. N., and Iversen, L. L.,** The anticonvulsant MK-801 is a potent N-methyl-D-aspartate antagonist, *Proc. Natl. Acad. Sci. U.S.A.*, 83, 7104, 1986.
51. **Engberg, I., Flatman, J. A., and Lambert, J. D. C.,** The action of N-methyl-D-aspartate and kainic acids on motoneurones with emphasis on conductance changes, *Br. J. Pharmacol.*, 64, 384P, 1978.
52. **Ault, B., Evans, R. H., Francis, A. A., Oakes, D. J., and Watkins, J. C.,** Selective depression of excitatory amino acid induced depolarizations by magnesium ions in isolated spinal cord preparations, *J. Physiol. (London)*, 307, 413, 1980.
53. **Johnston, D., Hablitz, J. J., and Wilson, W. A.,** Voltage clamp discloses slow inward current in hippocampal burst-firing neurones, *Nature (London)*, 286, 391, 1980.
54. **MacDonald, J. F. and Wojtowicz, J. M.,** Two conductance mechanisms activated by applications of L-glutamic, L-aspartic, DL-homocysteic, N-methyl-D-aspartic, and DL-kainic acids to cultured mammalian central neurons, *Can. J. Physiol. Pharmacol.*, 58, 1393, 1980.
55. **Gration, K. A. F., Lambert, J. J., Ramsey, R. L., Rand, R. P., and Usherwood, P. N. R.,** Agonist potency determination by patch clamp analysis of single glutamate receptors, *Brain Res.*, 230, 400, 1981.
56. **Hablitz, J. J.,** Conductance changes induced by DL-homocysteic acid and N-methyl-DL-aspartic acid in hippocampal neurons, *Brain Res.*, 247, 149, 1982.
57. **MacDonald, J. F. and Wojtowicz, J. M.,** The effects of L-glutamate and its analogues upon the membrane conductance of central murine neurones in culture, *Can. J. Physiol. Pharmacol.*, 60, 282, 1982.
58. **Collingridge, G. L. S., Kehl, S. J., and McLennan, H.,** Excitatory amino acids in synaptic transmission in the Schaffer collateral-commissural pathway of the rat hippocampus, *J. Physiol. (London)*, 334, 33, 1983.
59. **Dingledine, R.,** N-methyl aspartate activates voltage-dependent calcium conductance in rat hippocampal pyramidal cells, *J. Physiol. (London)*, 343, 385, 1983.
60. **Flatman, J. A., Schwindt, P. C., Crill, W. E., and Stafstrom, C. E.,** Multiple actions of N-methyl-D-aspartate on cat neocortical neurons in vitro, *Brain Res.*, 266, 169, 1983.

61. **Nistri, A. and Arenson, M. S.,** Differential sensitivity of spinal neurones to amino acids: an intracellular study on the frog spinal cord, *Neuroscience*, 8, 115, 1983.
62. **Yamamoto, C. and Takagi, M.,** Membrane potential fluctuations associated with glutamate- and homocysteate-induced depolarizations in hippocampal neurons, *Neurosci. Lett.*, 40, 257, 1983.
63. **Mayer, M. L., Westbrook, G. L., and Guthrie, P. B.,** Voltage-dependent block by Mg^{2+} of NMDA responses in spinal cord neurones, *Nature (London)*, 309, 261, 1984.
64. **Nowak, L. P., Bregestovski, P., Ascher, A. H., and Prochiantz, A.,** Magnesium gates glutamate-activated channels in mouse central neurones, *Nature (London)*, 307, 462, 1984.
65. **Coan, E. J. and Collingridge, G. L.,** Magnesium ions block an N-methyl-D-aspartate receptor-mediated component of synaptic transmission in rat hippocampus, *Neurosci. Lett.*, 53, 21, 1985.
66. **Hablitz, J. J.,** Action of excitatory amino acids and their antagonists on hippocampal neurons, *Cell. Mol. Neurobiol.*, 5, 389, 1985.
67. **Yamamoto, C. and Sato, H.,** Some properties of ionic channels activated by excitatory amino acids in hippocampal neurons, *Exp. Brain Res.*, 57, 313, 1985.
68. **Herron, C. E., Lester, R. A. J., Coan, E. J., and Collingridge, G. L.,** Frequency-dependent involvement of NMDA receptors in the hippocampus: a novel synaptic mechanism, *Nature (London)*, 322, 265, 1986.
69. **MacDermott, A. B., Mayer, M. L., Westbrook, G. L., Smith, S. J., and Barker, J. L.,** NMDA-receptor activation increases cytoplasmic calcium concentration in cultured spinal cord neurones, *Nature (London)*, 323, 519, 1986.
70. **Collins, G. G.,** Some effects of excitatory amino acid receptor antagonists on synaptic transmission in the rat olfactory cortex slice, *Brain Res.*, 244, 311, 1982.
71. **Koerner, J. F. and Cotman, C. W.,** Response of Schaffer collateral-CA1 pyramidal cell synapses of the hippocampus to analogues of acidic amino acids, *Brain Res.*, 251, 105, 1982.
72. **Davies, J. and Watkins, J. C.,** Role of excitatory amino acid receptors in mono- and polysynaptic excitation in the cat spinal cord, *Exp. Brain. Res.*, 49, 280, 1983.
73. **Fagg, G. E. and Foster, A. C.,** Amino acid neurotransmitters and their pathway in the mammalian central nervous system, *Neuroscience*, 9, 701, 1983.
74. **Herron, C. E., Lester, R. A., Coan, E. J., and Collingridge, G. L.,** Intracellular demonstration of an N-methyl-D-aspartate receptor mediated component of synaptic transmission in the rat hippocampus, *Neurosci. Lett.*, 60, 19, 1985.
75. **Herrling, P. L.,** Pharmacology of the corticocaudate excitatory postsynaptic potential in the cat: evidence for its mediation by quisqualate- or kainate-receptors, *Neuroscience*, 14, 417, 1985.
76. **Jackson, H., Nemeth, E. F., and Parks, T. N.,** Non-N-methyl-D-aspartate receptors mediating synaptic transmission in the avian cochlear nucleus: effects of kynurenic acid, dipicolinic acid and streptomycin, *Neuroscience*, 16, 171, 1985.
77. **Davies, J., Miller, A. J., and Sheardown, M. J.,** Amino acid receptor mediated excitatory synaptic transmission in the cat red nucleus, *J. Physiol. (London)*, 376, 13, 1986.
78. **Salt, T. E.,** Mediation of thalamic sensory input by both NMDA receptors and non-NMDA receptors, *Nature (London)*, 322, 263, 1986.
79. **Thompson, A. M.,** A magnesium-sensitive post-synaptic potential in rat cerebral cortex resembles neuronal responses to N-methylaspartate, *J. Physiol. (London)*, 370, 531, 1986.
80. **Dunwiddie, T. and Lynch, G.,** Long-term potentiation and depression of synaptic responses in the rat hippocampus: localization and frequency dependency, *J. Physiol. (London)*, 276, 353, 1978.
81. **Collingridge, G. L., Kehl, S. J., and McLennan, H.,** Excitatory amino acids in synaptic transmission in the Schaffer-commissural pathway of the rat hippocampus, *J. Physiol. (London)*, 334, 33, 1983.
82. **Bliss, T. V. P. and Dolphin, A. C.,** What is the mechanism of long-term potentiation in the hippocampus?, *Trends Neurosci.*, 5, 289, 1982.
83. **Wigström, H. and Gustafson, B.,** A possible candidate of the postsynaptic condition for long-lasting potentiation in the guinea pig hippocampus in vitro, *Neurosci. Lett.*, 44, 327, 1984.
84. **Harris, E. W., Ganong, A. H., and Cotman, C. W.,** Long-term potentiation in the hippocampus involves activation of N-methyl-D-aspartate receptors, *Brain Res.*, 323, 132, 1984.
85. **Baxter, D. A., Bittner, G. D., and Brown, T. H.,** Quantal mechanism of long-term synaptic potentiation, *Proc. Natl. Acad. Sci. U.S.A.*, 82, 5978, 1985.
86. **Wigström, H., Gustafsson, B., and Huang, Y. Y.,** Mode of action of excitatory amino acid receptor antagonists on hippocampal long-lasting potentiation, *Neuroscience*, 17, 1105, 1986.
87. **Morris, R. G. M., Anderson, E., Lynch, G. S., and Baudry, M.,** Selective impairment of learning and blockade of long-term potentiation by an N-methyl-D-aspartate receptor antagonist, AP5, *Nature (London)*, 319, 774, 1986.
88. **Herrling, P. L., Morris, R., and Salt, T. E.,** Effects of excitatory amino acid and their antagonists on membrane and action potentials of cat caudate neurones, *J. Physiol. (London)*, 339, 207, 1983.
89. **Herron, C. E., Williamson, R., and Collingridge, G. L.,** A selective N-methyl-D-aspartate antagonist depresses epileptiform activity in rat hippocampal slices, *Neurosci. Lett.*, 61, 255, 1985.

90. **Luini, A., Goldberg, O., and Teichberg, V. I.,** Distinct pharmacological properties of excitatory amino acid receptors in rat striatum: study by a novel Na^+ efflux assay, *Proc. Natl. Acad. Sci. U.S.A.*, 78, 3250, 1981.
91. **Heinemann, U. and Pumain, R.,** Extracellular calcium activity changes in cat sensorimotor cortex induced by iontophoretic application of amino acids, *Exp. Brain Res.*, 40, 247, 1980.
92. **Zanatto, L. and Heinemann, U.,** Aspartate and glutamate induced reductions in extracellular free calcium and sodium concentration in area CA1 of "in vitro" hippocampal slices of rats, *Neurosci. Lett.*, 35, 79, 1983.
93. **Berdichevsky, E., Riveros, N., Sánchez-Armáss, S., and Orrego, F.,** Kainate, N-methylaspartate and other excitatory amino acids increase calcium influx into rat brain cortex cells in vitro, *Neurosci. Lett.*, 36, 75, 1983.
94. **Wroblewski, J. T., Nicoletti, F., and Costa, E.,** Different coupling of excitatory amino acid receptors with Ca^{2+} channels in primary cultures of cerebellar granule cells, *Neuropharmacology*, 24, 919, 1985.
95. **Riveros, N. and Orrego, F.,** N-Methylaspartate-activated calcium channels in rat brain cortex slices, effect of calcium channel blockers and of inhibitory and depressant substances, *Neuroscience*, 17, 541, 1986.
96. **Miller, S. G. and Kennedy, M. B.,** Regulation of brain. II. Ca^{2+}/calmodulin-dependent protein kinase by autophosphorylation: a Ca^{2+}-triggered molecular switch, *Cell*, 44, 861, 1986.
97. **Saitoh, T. and Schwartz, J. H.,** Phosphorylation of a Ca^{2+}/calmodulin-dependent protein kinase produces an autonomous enzyme in aplysia neurons, *J. Cell Biol.*, 100, 835, 1985.
98. **Siesjö, B. K.,** Calcium and ischemic brain damage, *Eur. Neurol.*, 25, 45, 1986.
99. **Foster, G. A. and Roberts, P. J.,** Stimulation of rat cerebellar guanosine 3',5'-cyclic monophosphate (cyclic GMP) levels: effects of amino acid antagonists, *Br. J. Pharmacol.*, 74, 723, 1981.
100. **Novelli, A. and Henneberry, R. C.,** Veratrine- and glutamate-stimulated cGMP production in cultured rat neurons is differentially inhibited by imipramine, *Eur. J. Pharmacol.*, 118, 189, 1985.
101. **Sladeczek, F., Pin, J.-P., Récasens, M., Bockaert, J., and Weiss, S.,** Glutamate stimulates inositol phosphate formation in striatal neurones, *Nature (London)*, 317, 717, 1985.
102. **Baudry, M., Evans, J., and Lynch, G.,** Excitatory amino acids inhibit stimulation of phosphatidylinositol metabolism by aminergic agonists in hippocampus, *Nature (London)*, 319, 329, 1986.
103. **Nicoletti, F., Wroblewski, J. T., Novelli, A., Alho, H., Guidotti, A., and Costa, E.,** The activation of inositol phospholipid metabolism as a signal-transducing system for excitatory amino acids in primary cultures of cerebellar granule cells, *J. Neurosci.*, 6, 1905, 1986.
104. **Scatton, B. and Lehmann, J.,** N-methyl-D-aspartate-type receptors mediate striatal ^{3}H-acetyl choline release evoked by excitatory amino acids, *Nature (London)*, 297, 422, 1982.
105. **Drejer, J., Honoré, T., and Schousboe, A.,** Excitatory amino acid induced release of ^{3}H-GABA from cultured mouse cerebral cortex interneurons, *J. Neurosci.*, 7, 2910, 1987.
106. **Michaelis, E. K., Michaelis, M. L., and Boyarski, L. L.,** High-affinity glutamate binding to brain synaptic membranes, *Biochim. Biophys. Acta*, 367, 338, 1974.
107. **Roberts, P. J.,** Glutamate receptors in rat CNS, *Nature (London)*, 252, 399, 1974.
108. **Foster, A. C. and Roberts, P. J.,** High affinity L-(^{3}H)glutamate binding to postsynaptic sites on rat cerebellar membranes, *J. Neurochem.*, 31, 1467, 1978.
109. **Biziere, K., Thompson, H., and Coyle, J. T.,** Characterization of specific, high-affinity binding sites for L-(^{3}H)-glutamic acid in rat membranes, *Brain Res.*, 183, 421, 1980.
110. **de Barry, J., Vincendon, G., and Gombos, G.,** High-affinity glutamate binding during postnatal development of rat cerebellum, *FEBS Lett.*, 109, 175, 1980.
111. **Head, R. A., Tunnicliff, G., and Matheson, G. K.,** Glutamate receptor binding to cat central nervous system membranes, *Can. J. Biochem.*, 58, 534, 1980.
112. **Baudry, M. and Lynch, G.,** Characterization of two (^{3}H)glutamate binding sites in rat hippocampal membranes, *J. Neurochem.*, 36, 811, 1981.
113. **Foster, A. C., Mena, E. E., Fagg, G. E., and Cotman, C. W.,** Glutamate and aspartate binding sites are enriched in synaptic junctions isolated from rat brain, *J. Neurosci.*, 1, 620, 1981.
114. **Honoré, T., Lauridsen, J., and Krogsgaard-Larsen, P.,** Ibotenic acid analogues as inhibitors of (^{3}H)glutamic acid binding to cerebellar membranes, *J. Neurochem.*, 36, 1302, 1981.
115. **Michaelis, E. K., Michaelis, M. L., Chang, H. H., Grubs, R. D., and Kuonen, D. R.,** Molecular characteristics of glutamate receptors in the mammalian brain, *Mol. Cell. Biochem.*, 38, 163, 1981.
116. **Sharif, N. A. and Roberts, P. J.,** Regulation of cerebellar L-^{3}H-glutamate binding: influence of guanine nucleotides and Na^+ ions, *Biochem. Pharmacol.*, 30, 3019, 1981.
117. **Fagg, G. E., Foster, A. C., Mena, E. E., and Cotman, C. W.,** Chloride and calcium ions reveal a pharmacologically-distinct population of L-glutamate binding sites in synaptic membranes: correspondence between biochemical and electrophysiological data, *J. Neurosci.*, 2, 958, 1982.
118. **Larder, A. P. and McLennan, H.,** The effect of Ca^{2+} on the pharmacology of (^{3}H)glutamate binding sites in rat hippocampus, *Neurosci. Lett.*, 32, 59, 1982.

119. **Sanderson, C. and Murphy, S.,** Glutamate binding in the rat cerebral cortex during ontogeny, *Dev. Brain Res.*, 2, 329, 1982.
120. **Slevin, J., Collins, J. F., Lindsley, K., and Coyle, J. T.,** Specific binding of (^{3}H)glutamate to cerebellar membranes: evidence for recognition site heterogeneity, *Brain Res.*, 249, 353, 1982.
121. **Werling, L. L., Mena, E. E., and Nadler, J. V.,** Complex binding of L-(^{3}H)glutamate to hippocampal synaptic membranes in the absence of sodium, *J. Neurochem.*, 38, 1050, 1982.
122. **Fagg, G. E., Monaghan, D. T., and Cotman, C. W.,** Freezing eliminates a specific population of L-glutamate receptors in synaptic membranes, *Neurosci. Lett.*, 38, 157, 1983.
123. **Fagg, G. E., Riederer, B., and Matus, A.,** Sodium ions regulate a specific population of acidic amino acid receptors in synaptic membranes, *Life Sci.*, 34, 1739, 1984.
124. **Baudry, M., Kramer, K., and Lynch, G.,** Irreversibility and time course of calcium stimulated (^{3}H)glutamate binding to rat hippocampal membranes, *Brain Res.*, 270, 142, 1983.
125. **Fagg, G. E. and Matus, A. I.,** Selective association of N-methyl-aspartate and quisqualate types of L-glutamate receptor with brain postsynaptic densities, *Proc. Natl. Acad. Sci. U.S.A.*, 81, 6876, 1984.
126. **Wu, K., Carlin, R., and Siekevitz, P.,** Binding of L-(^{3}H)glutamate to fresh or frozen synaptic membrane and postsynaptic density fractions isolated from cerebral cortex and cerebellum of fresh or frozen canine brain, *J. Neurochem.*, 46, 831, 1986.
127. **Foster, A. C. and Fagg, G. E.,** Acidic amino acid binding sites in mammalian neuronal membranes: their characteristics and relationship to synaptic receptors, *Brain Res. Rev.*, 7, 103, 1984.
128. **Balcar, V. J. and Johnston, G. A. R.,** Structural specificity of high affinity uptake of L-glutamate and L-aspartate by rat brain slices, *J. Neurochem.*, 19, 2657, 1972.
129. **Vincent, S. R. and McGeer, E. G.,** A comparison of sodium-dependent glutamate binding with high-affinity glutamate uptake in rat striatum, *Brain Res.*, 184, 99, 1980.
130. **Cohen, R. S., Blomberg, F., Berzins, K., and Siekevitz, P.,** The structure of postsynaptic densities isolated from dog cerebral cortex. I. Overall morphology and protein composition, *J. Cell Biol.*, 74, 181, 1977.
131. **Fagg, G. E. and Lanthorn, T. H.,** Cl^-/Ca^{2+}-dependent L-glutamate binding sites do not correspond to 2-amino-4-phosphonobutanoate-sensitive excitatory amino acid receptors, *Br. J. Pharmacol.*, 86, 743, 1985.
132. **Pin, J.-P., Bockaert, J., and Recasens, M.,** The Ca^{2+}/Cl^--dependent L-(^{3}H)glutamate binding: a new receptor or a particular transport process?, *FEBS Lett.*, 175, 1774, 1984.
133. **Danbolt, N. C. and Storm-Mathisen, J.,** Na^+-dependent "binding" of D-aspartate in brain membranes and its inhibition by K^+, *J. Neurochem.*, 47, 819, 1986.
134. **Honoré, T., Drejer, J., Nielsen, M., and Braestrup, C.,** Differentiation of Cl^-/Ca^{2+}-dependent and sodium-dependent ^{3}H-glutamate binding to cortical membranes from rat brain by high energy radiation inactivation analysis, *J. Neural Transm.*, 65, 93, 1986.
135. **Snodgrass, S. R.,** In vitro binding studies with ^{3}H-N-methyl-aspartate, *Soc. Neurosci. Abstr.*, 5, 1943, 1979.
136. **Jones, A. W., Olwerman, H. J., and Watkins, J. C.,** A study of N-methyl-D-aspartate receptor sites on rat brain membranes using (^{3}H)D(−)-2-amino-5-phosphonovalerate as a labelled ligand, *J. Physiol. (London)*, 340, 45P, 1983.
137. **Ferkany, J. W. and Coyle, J. T.,** Specific binding of (^{3}H)-(+)-2-amino-7-phosphonoheptanoic acid to rat brain membranes in vitro, *Life Sci.*, 33, 1295, 1983.
138. **Monaghan, D. T., Yao, D., Olverman, H. J., Watkins, J. C., and Cotman, C. W.,** Autoradiography of D-(^{3}H)-2-amino-5-phosphonopentanoate binding sites in rat brain, *Neurosci. Lett.*, 52, 253, 1984.
139. **Murphy, D. E., Schneider, J., Lehmann, J., and Williams, M.,** Binding of (^{3}H)CPP (3(2-carboxypi-perazin-4-yl)propyl-1-phosphonic acid): a selective, high affinity ligand for N-methyl-D-aspartate(NMDA/EA-1) receptors, *Soc. Neurosci. Abstr.*, 12, 21.4, 1986.
140. **Fagg, G. E. and Baud, J.,** Activation of N-methyl-D-aspartate (NMDA) receptors unmasks ^{3}H-TCP binding sites in brain membranes: further evidence that phencyclidine blocks the open NMDA receptor channel, *Soc. Neurosci. Abstr.*, 12, 264.6, 1986.
141. **Loo, P., Brunwalder, A., Lehmann, J., and Williams, M.,** Radioligand binding to central phencyclidine recognition sites is dependent on excitatory amino acid receptor agonists, *Eur. J. Pharmacol.*, 123, 467, 1986.
142. **Honoré, T., Lauridsen, J., and Krogsgaard-Larsen, P.,** The binding of (^{3}H)AMPA, a structural analogue of glutamic acid, to rat brain membranes, *J. Neurochem.*, 38, 173, 1982.
143. **Hansen, J. J., Lauridsen, J., Nielsen, E., and Krogsgaard-Larsen, P.,** Enzyme resolution and binding to rat brain membranes of the glutamic acid agonist α-amino-3-hydroxy-5-methyl-4-isoxazolepropionic acid, *J. Med. Chem.*, 26, 901, 1983.
144. **Honoré, T. and Nielsen, M.,** Complex structure of quisqualate-sensitive glutamate receptors in rat cortex, *Neurosci. Lett.*, 54, 27, 1986.
145. **Honoré, T. and Drejer, J.,** Chaotropic ions change the conformation of quisqualate receptors, *Soc. Neurosci. Abstr.*, 16, 107.12, 1986.

146. **Simon, J. R., Contrera, J. F., and Kuhar, M. J.,** Binding of (^{3}H)-kainic acid, an analogue of L-glutamate to brain membranes, *J. Neurochem.*, 26, 141, 1976.
147. **Vincent, S. R. and McGeer, E. G.,** Kainic acid binding to membranes of striatal neurons, *Life Sci.*, 24, 265, 1979.
148. **London, E. D. and Coyle, J. T.,** Specific binding of ^{3}H-kainic acid to receptor sites in rat brain, *Mol. Pharmacol.*, 15, 492, 1979.
149. **Nieto-Sampedro, M., Shelton, D., and Cotman, C. W.,** Specific binding of kainic acid to purified subcellular fractions from rat brain, *Neurochem. Res.*, 5, 591, 1980.
150. **Foster, A. C., Mena, E. E., Monaghan, D. T., and Cotman, C. W.,** Synaptic localization of kainic acid binding sites, *Nature (London)*, 289, 73, 1981.
151. **Slevin, J. T., Collins, J. F., and Coyle, J. T.,** Analogue interactions with the brain receptor labelled by (^{3}H)kainic acid, *Brain Res.*, 265, 169, 1983.
152. **Honoré, T., Drejer, J., and Nielsen, M.,** Calcium discriminates two (^{3}H)kainate binding sites with different molecular target sizes in rat cortex, *Neurosci. Lett.*, 65, 47, 1986.
153. **Foster, A. C., Fagg, G. E., Mena, E. E., and Cotman, C. W.,** L-Glutamate and L-aspartate bind to separate sites in rat brain synaptic membranes, *Brain Res.*, 229, 246, 1981.
154. **Sharif, N. A. and Roberts, P. J.,** L-Aspartate binding sites in rat cerebellum: a comparison of the binding of L-(^{3}H)aspartate and L-(^{3}H)glutamate to synaptic membranes, *Brain Res.*, 211, 293, 1981.
155. **Mitchell, C. K. and Redburn, D. A.,** 2-Amino-4-phosphonobutyric acid and N-methyl-D-aspartate differentiate between (^{3}H)glutamate and (^{3}H)aspartate binding sites in bovine retina, *Neurosci. Lett.*, 28, 241, 1982.
156. **DiLauro, A., Meek, J. L., and Costa, E.,** Specific high-affinity binding of L-^{3}H-aspartate to rat brain membranes, *J. Neurochem.*, 38, 1261, 1982.
157. **Butcher, S. P., Collins, J. F., and Roberts, P. J.,** Characterization of the binding of DL-(^{3}H)2-amino-4-phosphonobutyrate to L-glutamate-sensitive sites on rat brain synaptic membranes, *Br. J. Pharmacol.*, 80, 355, 1983.
158. **Monaghan, D. T., McMills, M. C., Chamberlain, A. R., and Cotman, C. W.,** Synthesis of (^{3}H)2-amino-4-phosphonobutyric acid and characterization of its binding to rat brain membranes: a selective ligand for the chloride/calcium-dependent class of L-glutamate binding sites, *Brain Res.*, 278, 137, 1983.
159. **Iwata, H., Yamagami, S., and Baba, A.,** Cysteine sulphinic acid in the central nervous system: specific binding of (^{35}S)cysteic acid to cortical synaptic membranes — an investigation of possible binding sites for cysteine sulphinic acid, *J. Neurochem.*, 38, 1275, 1982.
160. **Recasens, M., Varga, V., Nanopoulos, D., Saadoun, F., Vincendon, G., and Benavides, J.,** Evidence for cysteine sulfinate as a neurotransmitter, *Brain Res.*, 239, 153, 1982.
161. **Greenamyre, J. T., Young, A. B., and Penney, J. B.,** Quantitative autoradiography of L-(^{3}H)glutamate binding to rat brain, *Neurosci. Lett.*, 37, 155, 1983.
162. **Halpain, S., Parsons, B., and Rainbow, T. C.,** Tritium-film autoradiography of sodium-independent glutamate binding sites in rat brain, *Eur. J. Pharmacol.*, 86, 313, 1983.
163. **Monaghan, D. T., Holets, V. R., Toy, D. W., and Cotman, C. W.,** Anatomical distributions of four pharmacologically distinct ^{3}H-L-glutamate binding sites, *Nature (London)*, 306, 176, 1983.
164. **Halpain, S., Wieczorek, C. M., and Rainbow, T. C.,** Localization of L-glutamate receptors in rat brain by quantitative autoradiography, *J. Neurosci.*, 4, 2247, 1984.
165. **Greenamyre, J. T., Young, A. B., and Penney, J. B.,** Quantitative autoradiographic distribution of L-(^{3}H) glutamate-binding sites in rat central nervous system, *J. Neurosci.*, 4, 2133, 1984.
166. **Greenamyre, J. T., Olson, J. M. M., Penney, J. B., and Young, A. B.,** Autoradiographic characterization of N-methyl-D-aspartate-, quisqualate- and kainate-sensitive glutamate binding sites, *J. Pharmacol. Exp. Ther.*, 233, 254, 1985.
167. **Monaghan, D. T. and Cotman, C. W.,** Distribution of N-methyl-D-aspartate-sensitive L-(^{3}H)glutamate-binding sites in rat brain, *J. Neurosci.*, 5, 2909, 1985.
168. **Maragos, W. F., Chu, D. C. M., Greenamyre, J. T., Penney, J. B., and Young, A. B.,** High correlation between the localization of (^{3}H)TCP binding and NMDA receptors, *Eur. J. Pharmacol.*, 123, 173, 1986.
169. **Hösli, E., Krogsgaard-Larsen, P., and Hosli, L.,** Binding sites for the glutamate analogue (^{3}H)AMPA in cultured rat brainstem and spinal cord, *Brain Res.*, 268, 177, 1983.
170. **Monaghan, D. T., Yao, D., and Cotman, C. W.,** Distribution of (^{3}H)AMPA binding sites in rat brain as determined by quantitative autoradiography, *Brain Res.*, 324, 160, 1984.
171. **Rainbow, T. C., Wieczorek, C. M., and Halpain, S.,** Quantitative autoradiography of binding sites for (^{3}H)-AMPA, a structural analogue of glutamic acid, *Brain Res.*, 309, 173, 1984.
172. **Monaghan, D. T. and Cotman, C. W.,** The distribution of (^{3}H)kainic acid binding sites in rat CNS as determined by autoradiography, *Brain Res.*, 252, 91, 1982.
173. **Unnerstall, J. R. and Walmsley, J. K.,** Autoradiographic localization of high-affinity (^{3}H)kainic acid binding sites in rat forebrain, *Eur. J. Pharmacol.*, 86, 361, 1983.

174. **Monaghan, D. T., Nguyen, L., and Cotman, C. W.,** The distribution of (^{3}H)kainate binding sites in primate hippocampus is similar to the distribution of both Ca^{2+}-sensitive and Ca^{2+}-insensitive (^{3}H)kainate binding sites in rat hippocampus, *Neurochem. Res.*, 11, 1073, 1986.
175. **Slaughter, M. M. and Miller, R. F.,** 2-Amino-4-phosphonobutyric acid: a new pharmacological tool for retina research, *Science*, 211, 182, 1981.
176. **Slaughter, M. M. and Miller, R. F.,** The role of excitatory amino acid transmitters in the mudpuppy retina an analysis with kainic acid and N-methyl aspartate, *J. Neurosci.*, 3, 1701, 1983.
177. **Koerner, J. F. and Cotman, C. W.,** Micromolar L-2-amino-4-phosphonobutyric acid selectively inhibits perforant path synapses from the lateral entorhinal cortex, *Brain Res.*, 216, 192, 1981.
178. **Yamamoto, C., Sawada, S., and Takada, S.,** Suppressing action of 2-amino-4-phosphonobutyric acid on mossy fiber-induced excitation in the guinea pig hippocampus, *Exp. Brain Res.*, 51, 128, 1983.
179. **Koerner, J. F., Whittemore, E. R., and Robinson, M. B.,** Quisqualate sensitizes synaptic responses in hippocampal slices to phosphonate-containing analogues of glutamate, *Soc. Neurosci. Abstr.*, 12, 93.13, 1986.
180. **Werling, L. L., Doman, A., and Nadler, J. V.,** L-(^{3}H)glutamate binding to hippocampal synaptic membranes: two binding sites discriminated by their differing affinities for quisqualate, *J. Neurochem.*, 41, 586, 1983.
181. **Nadler, J. V., Wang, A., and Werling, L. L.,** Binding sites for L-(^{3}H)glutamate on hippocampal synaptic membranes: three populations differentially affected by chloride and calcium ions, *J. Neurochem.*, 44, 1791, 1985.
182. **Drejer, J., Honoré, T., Meier, E., and Schousboe, A.,** Pharmacologically distinct glutamate receptors on cerebellar granule cells, *Life Sci.*, 38, 2077, 1986.
183. **Erez, U., Frenk, H., Goldberg, O., Cohen, A., and Teichberg, V. I.,** Anticonvulsant properties of 3-hydroxy-2-quinoxaline-carboxylic acid, a newly found antagonist of excitatory amino acids, *Eur. J. Pharmacol.*, 110, 31, 1985.
184. **Watkins, J. C.,** personal communication.

Pathology of Glutamine and Glutamate

Chapter 21

GLUTAMINE, GLUTAMATE, AND GABA IN HUMAN DISEASES

Thomas L. Perry

TABLE OF CONTENTS

I. INTRODUCTION

Large amounts of glutamine, glutamate, and GABA are present in the human brain, and both plasma and cerebrospinal fluid (CSF) contain high concentrations of glutamine. Plasma contains readily measurable concentrations of glutamate, and much smaller concentrations of glutamate and GABA are present in human CSF. Since GABA and glutamic acid are important as inhibitory and excitatory neurotransmitters, respectively, in the central nervous system (CNS) and since glutamine constitutes a nontoxic storage form for ammonia, accurate determination of these three compounds in physiological fluids and tissues is of great importance in studies of human diseases, especially those involving the CNS. Conventional automatic amino acid analyses of fasting plasma and CSF, using ninhydrin for detection, have provided many clues to amino acid abnormalities of the brain in various metabolic and neurological disorders. Equally useful has been the direct measurement of amino acid contents in autopsied or biopsied human brain by this technique. More recently, the development of greatly enhanced sensitivity in detection of amino compounds, using the fluorescence produced by their reaction with *o*-phthalaldehyde, has made it possible to quantitate accurately the very low levels of GABA in CSF.

II. NORMAL GLUTAMINE, GLUTAMATE, AND GABA LEVELS IN HUMANS

Interpretation of the significance of altered levels of glutamine, glutamate, or GABA in human diseases requires reliable information as to what the normal values are during life and how these values may be altered in tissues following death. While most values reported for glutamine in the literature in recent years appear reasonably accurate, there have been many grossly unreliable values reported for the concentrations of glutamate and GABA in CSF.

If excessive amounts of sulfosalicyclic acid are added to CSF as a deproteinizing agent prior to its application to the cation exchange column of an automatic amino acid analyzer, the upper layers of the cation exchange resin are converted by the excess acid into an excellent hydrolyzing medium, especially if the analyzer column is operated at a high temperature.[1] This will then cause some of the glutamine present in the CSF to be hydrolyzed to form artefactual glutamate and will cause some of the homocarnosine (γ-aminobutyryl-histidine) present in CSF to be hydrolyzed to form artefactual GABA. The same mechanism can cause on-column hydrolysis of glutamine to glutamate (and of asparagine to aspartate) in plasma samples deproteinized with excessive sulfosalicylic acid. This technical error is especially likely to occur with modern amino acid analyzers which utilize columns with a very small internal diameter and which therefore contain relatively small amounts of cation exchange resin. In addition, if an amino acid analyzer is operated at a column temperature higher than 35°C prior to the elution of glutamine, part of this compound can be cyclized to form the ninhydrin-negative compound pyrrolidone carboxylic acid.[2] This technical error can result in unduly low values for glutamine in plasma and CSF.

In studies of plasma or CSF, contamination of the fluid with cells, especially leukocytes and platelets, can lead to artefactually high values of glutamate and of three other amino compounds which are present in high concentrations intracellularly, namely, taurine, phosphoethanolamine, and aspartic acid.[2] Thus, glutamate values are apt to be artefactually high if CSF is analyzed after a traumatic lumbar puncture or when blood has been centrifuged at a relatively slow speed and the resulting platelet-rich plasma is deproteinized and utilized for amino acid analysis.

Table 1 presents normal mean values for glutamine, glutamic acid, and GABA, as determined in my laboratory over a number of years, in human plasma, CSF, two regions of biopsied brain, and erythrocytes. The mean values shown for glutamine and glutamate in

Table 1
NORMAL VALUES FOR GLUTAMINE, GLUTAMATE, AND GABA IN HUMANS

Fluid or tissue	Age	Glutamine	Glutamic acid	GABA
Fasting plasma	Adults[a]	624 ± 118 (89)	23 ± 11 (89)	
	Children (<5 years)	523 ± 105 (53)		
CSF	Adults[b]	587 ± 86 (88)	0.7 ± 0.6 (190)	84 ± 36 (35)
	Children (<5 years)	499 ± 137 (45)		
	Children (birth—10 years)		0.9 ± 0.8 (64)	
Cerebral cortex (biopsied)	Adults	5.42 ± 1.84 (14)	7.58 ± 1.86 (14)	0.63 ± 0.21 (14)
Cerebellar cortex (biopsied)	Adults	7.26 ± 1.15 (9)	7.58 ± 1.42 (9)	0.74 ± 0.13 (9)
Erythrocytes	Adults	0.34 ± 0.10 (18)	0.18 ± 0.05 (18)	

Note: Mean values for glutamine and glutamate in plasma and CSF are expressed in μmol/ℓ, and for GABA they are expressed in nmol/ℓ. Mean contents of amino acids in biopsied brain and erythrocytes are expressed in μmol/g wet weight. Number of subjects examined is shown in parentheses.

[a] Healthy subjects.
[b] Subjects without neurological or psychiatric disorder.

plasma were obtained from 89 healthy adults. Mean plasma glutamine and glutamate values for infants and children were obtained both from well and sick children, but none of the sick patients had any disorder known to involve the metabolism of these two amino acids. Glutamine and glutamate mean concentrations in CSF were calculated chiefly from patients with neurological or psychiatric disorders, since in practice it has been difficult to obtain CSF from patients without any CNS disorder. The CSF glutamine value shown for adults in Table 1 is similar to the value that we published a decade ago,[3] but the mean value for glutamate in CSF is substantially lower than our earlier published mean.[3] A recent compilation of the glutamate concentrations in CSF specimens from 190 adults, specimens which we feel were obtained and analyzed under excellent conditions, has cut in half the glutamate value which we earlier considered normal for CSF.

Values shown in Table 1 for the glutamine, glutamate, and GABA contents of living human brain were obtained from brain specimens frozen in liquid nitrogen within 10 to 30 sec of occluding the blood supply.[4] These specimens of brain were obtained chiefly from patients undergoing removal of a deep-seated tumor or abscess, in which case it was necessary to sacrifice overlying cortical tissue. Some of the specimens of cerebral cortex (all were from frontal or temporal cortex) were obtained during brain biopsies as part of the investigation of undiagnosed neurological illnesses. Thus, the values for cerebral cortex and cerebellar cortex shown in Table 1 were all obtained from ill patients, but none of them had disorders known to affect glutamine, glutamic acid, or GABA metabolism. GABA content in the mammalian brain rises very rapidly after death,[5,6] and specimens must be frozen within seconds after the blood supply is cut if GABA values approximating those present during life are to be obtained. The mean values for glutamine, glutamic acid, and GABA presented in Table 1 are probably as close an approximation as can be obtained to the real contents of these three compounds in the superficial areas of living human brain.

The mean contents of glutamine and glutamate shown in Table 1 for erythrocytes were obtained from 18 adults, 8 of whom were healthy controls, while the remaining 10 had metabolic disorders, none of them known to involve abnormalities of glutamine or glutamate metabolism. None of the subjects had any hematological disorder, and none were anemic or had increased reticulocyte counts. Fresh erythrocytes were obtained by centrifugation of heparinized venous blood and were washed twice with three volumes of ice-cold 0.9% saline solution to remove amino acids in plasma still present in the packed erythrocytes.[7]

The methods of preparing plasma or CSF for quantitative amino acid analysis of compounds readily quantitated by ninhydrin detection (including glutamine and glutamate) which have been used in my laboratory are described elsewhere,[2,3] as is the single column lithium citrate buffer elution system used in chromatography.[8] Both autopsy and biopsy specimens of human brain were homogenized and deproteinized with perchloric acid[6,9,10] prior to amino acid analysis, as were the washed human erythrocytes.[7]

The mean GABA concentration of 84 nmol/ℓ for CSF of normal adults shown in Table 1 is much lower than values averaging 220 to 240 nmol/ℓ which have been published by other investigators,[11-14] who have used ion-exchange column chromatography of CSF, followed by reaction of amino compounds in the column eluate with *o*-phthalaldehyde and measurement of the resulting fluorescence. The probable reasons for the inordinately high CSF GABA values obtained by others are explained in the description of our methodology[1] and primarily involve on-column hydrolysis of homocarnosine (present in CSF at 20 to 100 times the concentration of free GABA) to form artefactual GABA. These errors have resulted from the use by most investigators of at least a tenfold excess of sulfosalicylic acid in deproteinizing CSF prior to applying it to the ion-exchange column. It is likely that even with the methods used in my laboratory,[1] some chemically bound GABA is freed during column chromatography. It may well be that true GABA concentrations in human CSF are even lower than the mean value shown in Table 1.

No value is given for GABA concentration in fasting plasma. In limited attempts to detect GABA in plasma, using our ion-exchange fluorometric method,[1] we have routinely found large interfering chromatographic peaks, some of them probably derived from the heparin used to prevent clotting of the blood specimen, which are eluted from the column at the GABA locus, but which are not GABA. Using our ninhydrin automatic amino acid analyzer technique,[8] we have never seen detectable amounts of GABA (i.e., well below 1 μmol/ℓ) in any plasma from many hundreds of subjects, with the single exception of low but readily measureable concentrations of GABA in the plasma of an experimental subject soon after he ingested GABA in a dosage of 1 mmol/kg.[15]

It is likely that the mean concentrations of glutamine and glutamate shown for fasting plasma and CSF in Table 1 are reasonably accurate and that the methods of sample preparation and column chromatography have neither destroyed glutamine nor produced artefactual glutamate. The plasma concentrations for both compounds shown in Table 1 agree well with values obtained many years ago in a large and carefully executed study by Armstrong and Stave,[16] and our plasma glutamate mean is almost as low as that recently reported by Bos and Slump.[17] Investigators should view with suspicion the methodology employed in investigations of human disease where control plasma and CSF values for glutamate are much higher than those listed in Table 1.

III. PATHOLOGICAL ALTERATIONS OF GLUTAMINE

Glutamine concentrations in plasma and CSF, and glutamine content in the brain, are usually elevated, and sometimes very markedly so, in conditions where ammonia is no longer

Table 2
CONDITIONS IN WHICH GLUTAMINE LEVELS ARE ABNORMAL IN HUMAN PLASMA, CSF, AND BRAIN

Condition	Fasting plasma	CSF	Brain
Hepatic encephalopathy	↑	↑ ↑	↑ ↑
Urea cycle disorders	↑	↑ ↑	↑ ↑
Lysinuric protein intolerance	↑	?	?
Pyruvate carboxylase deficiency	↓	↓	↓ ↓
Antiepileptic drugs			
Phenobarbitone, primidone	↑	↑	?
Valproate	↑	↑	?

Note: ↑, ↓ : increased, decreased; ↑ ↑, ↓ ↓ : markedly increased, decreased; and ?: not yet determined.

converted into urea (Table 2). These may result either from structural or biochemical failure in the liver or as a result of any of the six genetically determined metabolic defects in the urea cycle.[18,19] In addition, plasma concentrations of glutamine are sometimes elevated in the autosomal recessive disorder lysinuric protein intolerance.[18,20]

Ammonia which has not been disposed of normally in the urea cycle is "detoxified" by conversion into glutamine. Ammonia is first reacted with α-ketoglutarate (2-oxoglutarate) derived from the tricarboxylic acid (TCA) cycle to form glutamate. This reaction is catalyzed by glutamate dehydrogenase. The second step, catalyzed by glutamine synthetase, reacts further ammonia with glutamate to form glutamine. In the genetically determined urea cycle metabolic defects, levels of ammonia and glutamine in physiological fluids and the brain are usually highest in patients with carbamoyl phosphate synthetase deficiency and ornithine carbamoyl transferase deficiency and lower in defects later in the urea cycle such as argininosuccinate lyase deficiency and arginase deficiency.[18]

Concentrations of glutamine in CSF in urea cycle disorders and in hepatic encephalopathy are often higher than they are in specimens of plasma which have been obtained at the same time,[21] and CSF glutamine concentrations can be elevated as much as two to five times the normal values (about 600 μmol/ℓ in adults and about 500 μmol/ℓ in infants). Glutamine contents of autopsied human brain, which are similar to those found in biopsied human brain,[6] vary from 4 to 6 μmol/g of wet weight in different brain regions.[4] In autopsied brains examined in my laboratory from patients dying with hepatic encephalopathy due to hepatitis or alcoholic cirrhosis and in brains of infants dying with ornithine carbamoyl transferase deficiency or argininosuccinate lyase deficiency,[22] glutamine contents in different brain regions have ranged from 10 to 28 μmol/g of wet weight.

Table 2 also shows one metabolic disorder in which glutamine levels are reduced. In pyruvate carboxylase deficiency, one of several genetically determined forms of severe lactic acidosis in infants, glutamine concentrations of plasma and CSF have been found moderately reduced and glutamine contents in the brain were markedly reduced.[23] In this disorder, glutamine stores may be consumed, especially in the brain, to form α-ketoglutarate for replacement of oxaloacetate which is in short supply as a result of the pyruvate carboxylase deficiency and to keep the TCA cycle running.

Glutamine concentrations are sometimes found to be elevated in the plasma and CSF of patients given certain anticonvulsant drugs (Table 2), especially in infants and children treated with phenobarbitone or primidone.[24] Valproic acid has frequently been reported to produce moderate hyperammonemia[25-28] and can raise glutamine concentrations in both plasma and CSF. It has been suggested that valproate interferes in the urea cycle either by inhibiting carbamoyl phosphate synthetase, the first enzyme in the urea cycle, or by blocking

Table 3
HUMAN DISORDERS WITH ALTERED GLUTAMATE LEVELS IN PLASMA, CSF, AND BRAIN

Disorders	Fasting plasma	CSF	Brain
Olivopontocerebellar atrophy (OPCA), recessively inherited, with glutamate dehydrogenase deficiency	↑	?	?
OPCAs, dominantly inherited	N	N	N or ↓ ↓ (Cerebellar cortex)
Ataxia telangiectasia			↓ ↓ (Cerebellar cortex)
Focal epilepsies	N	N	N or ↑ (Biopsied cerebral cortical foci)
Huntington's chorea	N	N	↓ (Caudate, putamen, occipital cortex)

Note: ↑, ↓: increased, decreased ↓ ↓: markedly decreased; ?: not yet determined; and N: normal.

the synthesis of *N*-acetylglutamate, the essential activator of this enzyme.[26-28] Whether or not glutamine content is elevated in brains in some patients treated with phenobarbitone or valproate and whether or not enough ammonia accumulates in the brain to produce harmful effects remains unknown.

IV. PATHOLOGICAL ALTERATIONS OF GLUTAMIC ACID

Table 3 lists human disorders in which the concentrations of glutamate in fasting plasma or its contents in certain brain regions are known to be abnormal. In one form of olivopontocerebellar atrophy (OPCA) which is inherited as an autosomal recessive, activity of glutamate dehydrogenase is deficient in leukocytes and cultured fibroblasts,[29,30] and plasma concentrations of α-ketoglutarate are decreased while plasma glutamate concentrations are increased. Presumably the enzyme deficiency in this form of OPCA is also present in the brain, and glutamate would be expected to accumulate in the brain and CSF, as well as in plasma. Whether this is the case has not yet been reported.

There are several different forms of OPCA which are inherited as autosomal dominants, each form showing different biochemical abnormalities in the brain.[31] In two different dominant cerebellar disorders, each of which has been classified clinically[32] as OPCA, type I, the content of glutamic acid in the cerebellar cortex is markedly reduced, while in a third form of OPCA, type I, glutamate content is normal in this brain region.[31] In dominantly inherited OPCA, type IV, glutamate contents are greatly reduced both in the cerebellar cortex and in other brain regions such as the frontal cortex and the caudate nucleus.[31] Glutamate content was also found to be greatly reduced in the cerebellar cortex of the single patient with ataxia telangiectasia whose autopsied brain has been studied neurochemically.[33] Glutamate contents are similar in biopsied and autopsied human brain,[6] and the low glutamate contents found in cerebellar cortex in the OPCAs and ataxia telangiectasia after death almost certainly reflect deficiencies which were present during life. Glutamate is believed to be the excitatory neurotransmitter of the granule cells in the cerebellar cortex,[34-36] and the glutamate deficiencies found in these cerebellar disorders reflect marked loss of these particular neurons.

When epileptogenic brain tissue is removed neurosurgically from patients with focal epilepsy, glutamate content is substantially increased in a minority of these biopsied foci.[37] In these patients, it is possible that excessive amounts of glutamate, acting locally as an excitatory neurotransmitter, may have been responsible for initiating focal seizures. The mean contents of glutamic acid are significantly reduced in the caudate nucleus, putamen, and occipital cortex, but not in other brain regions, of patients dying with Huntington's chorea. Whether the striatal glutamate deficiency in Huntington's chorea simply reflects the marked neuronal loss characteristic for this region or whether it may indicate abnormally great release of excitatory neurotransmitter by glutamatergic corticostriatal neurons[38] is unknown.

Glutamate concentrations have often been reported to be increased in the plasma or CSF of patients with hepatic encephalopathy or with biochemical errors in the urea cycle. In a limited number of such patients examined in my laboratory, we have *not* found glutamate elevated in these fluids in the presence of hyperammonemia and markedly elevated glutamine values. Hence, elevation of glutamate in hepatic encephalopathy is not listed in Table 3. The use of grossly excessive amounts of sulfosalicylic acid when deproteinizing plasma or CSF specimens containing high concentrations of glutamine probably accounts for such reports of high glutamate concentrations in these fluids.[21]

One group of German investigators[39,40] has reported that glutamate concentrations are significantly reduced in the CSF of patients with schizophrenia, and they suggest that dysfunction of glutamatergic neurons in the brain may be important in the etiology of the characteristic symptoms of the schizophrenias. In my laboratory, we have found no reductions in levels of glutamate in either the CSF or the autopsied brain of schizophrenic patients.[41] A recent compilation of our CSF glutamate levels from 24 adult schizophrenics gave a value of 0.6 ± 0.2 (mean $\pm$SEM), while 190 controls with a variety of neurological diseases had a mean value of 0.7 ± 0.6. These values do not differ significantly. The mean glutamate concentration of CSF found by Kim et al.[39] is more than an order of magnitude higher than it should be. This emphasizes again the importance of preparing and analyzing physiological fluids in a manner designed to prevent artefacts produced by hydrolysis of other compounds.

V. PATHOLOGICAL ALTERATIONS OF GABA

GABA content is abnormal in the brain in many human neurological and psychiatric disorders, as shown by quantitative measurements of this amino acid in the autopsied brain. GABA content increases very rapidly after brain death,[5,6] but when the human brain is frozen 1 to 2 hr or later after death, a stable maximum content of GABA has been reached.[6] The post-mortem rise in GABA content in the brain is presumably due to continued activity of the synthesizing enzyme, glutamic acid decarboxylase (GAD), with early failure of the degradative enzyme systems utilizing GABA-aminotransferase (GABA-T) and succinic semialdehyde dehydrogenase. One can reasonably compare brain GABA content between patients and controls with the usual minimal delay of at least 1 hr between death and autopsy. The GABA values found in autopsied brains are certainly higher than during life. In descending order, the highest regional GABA contents (4 to 10 μmol/g of wet weight) are found in autopsied brain in globus pallidus, substantia innominata, substantia nigra, and nucleus accumbens. To learn how much GABA is present in the living human brain, one must instantly freeze surgically removed biopsies, and this information is obviously only available for cerebral and cerebellar cortex (Table 1).[4]

Table 4 lists brain disorders in which GABA contents are either known to be or suspected of being abnormal in humans. Huntington's chorea was the first disorder in which brain GABA content was found to be significantly reduced.[42] GABA content is low (Table 5) in the caudate nucleus, putamen, globus pallidus, substantia nigra, and occipital cortex in Huntington's chorea, but not in frontal or cerebellar cortex nor in the dentate nucleus, a region where GABA content is characteristically high. The reduced GABA contents in the basal ganglia in Huntington's chorea are almost certainly due to a selective dropout of GABAergic interneurons in the striatum, as well as of striatopallidal and striatonigral GABAergic neurons. GABA concentrations are *not* significantly reduced in the CSF of Hun-

Table 4
DISORDERS WITH ABNORMAL GABA LEVELS IN CSF AND BRAIN

Disorder or condition	CSF	Brain	Regions involved
Huntington's chorea	N	↓↓ or ↓	Caudate nucleus, putamen, globus pallidus, substantia nigra, occipital cortex
Dialysis encephalopathy	N or ↓	↓↓ or ↓	Cerebral cortex, cerebellar cortex, caudate nucleus, thalamus
OPCAs, dominantly inherited	N	↓ or ↓↓	Cerebellar cortex, dentate nucleus
Ataxia telangiectasia		↓↓	Cerebellar cortex, dentate nucleus
Hallervorden-Spatz disease	N	↓↓	Globus pallidus, substantia nigra
Alzheimer's disease	N	↓↓ or ↓	Frontal and occipital cortex
Schizophrenias	N	N or ↓	Nucleus accumbens, thalamus
Tardive dyskinesia	?	?	Basal ganglia
Multiple sclerosis, with action tremor	N or ↓	?	
Parkinson's disease	N	↑	Putamen
Succinic semialdehyde dehydrogenase deficiency	?	?	
GABA-T inhibiting drugs (hydrazine, γ-vinyl GABA)	↑↑	↑↑	Many regions

Note: ↑, ↓: increased, decreased; ↑↑, ↓↓: markedly increased, decreased; ?: unknown; and N: normal.

tington's chorea patients,[1] despite published reports to the contrary.[43,44] In my laboratory, 35 neurologically normal adults had a CSF GABA concentration (mean ±SEM) of 84 ± 6 nmol/ℓ, while for 40 Huntington's chorea patients CSF GABA was 80 ± 8 nmol/ℓ. Thus, measurement of GABA concentrations in CSF is valueless for the diagnosis or preclinical detection of Huntington's chorea, although it may be useful in biochemical monitoring of drugs intended to increase brain GABA content, as shown later.

A second disorder in which important reductions in brain GABA content are present is dialysis encephalopathy.[45,46] This neurological syndrome is a frequently fatal complication of repeated hemodialysis in patients with chronic renal failure and is associated with elevated aluminum contents in the brain at autopsy. Aluminum presumably gains entrance to the brain and other tissues in such patients either from aluminum contaminating the dialysis fluid or as a result of absorption of aluminum from the GI tract. End-stage renal failure patients are commonly given large doses of aluminum gels by mouth to prevent excessive phosphate absorption.[45] Whether or not aluminum poisoning causes the brain GABA deficiency found in dialysis encephalopathy is still unknown. In any case, GABA contents are significantly reduced in dialysis encephalopathy in frontal and occipital cortex, cerebellar cortex, caudate nucleus, and thalamus. In this case, the GABA deficiencies are not accompanied by any histologically demonstrable losses of neurons. Table 5 contrasts the brain regions with reduced and normal GABA contents in control subjects, Huntington's choreics, and patients with dialysis encephalopathy. In the latter condition, CSF GABA concentrations are sometimes normal and sometimes markedly reduced.[45]

GABA content is moderately reduced in the cerebellar cortex and is often markedly reduced in the dentate nucleus, the dominantly inherited OPCAs,[31] and ataxia telangiectasia[33] (Table 4). In these disorders, the GABA deficiency probably is due to specific loss of Purkinje neurons and cerebellar cortical interneurons. GABA is almost certainly the inhibitory neurotransmitter of most of the Purkinje cells,[47-49] and it is probably the neurotransmitter of one or more of the three types of inhibitory interneurons in the cerebellar cortex: the basket, stellate, and Golgi cells.[48,49] GABA content is also markedly reduced in the globus pallidus and substantia nigra in Hallervorden-Spatz disease (Table 4), in this case because of the

Table 5
REGIONAL GABA CONTENTS IN AUTOPSIED HUMAN BRAIN

Patients	Frontal cortex	Occipital cortex	Cerebellar cortex	Caudate nucleus	Putamen	Globus pallidus	Substantia nigra	Nucleus accumbens	Thalamus, medial-dorsal	Dentate nucleus
Controls	1.67 ± 0.09	1.83 ± 0.11	1.65 ± 0.08	2.90 ± 0.14	2.91 ± 0.20	7.32 ± 0.40	6.05 ± 0.26	4.15 ± 0.29	2.05 ± 0.12	4.77 ± 0.23
	(26)	(25)	(26)	(32)	(19)	(16)	(28)	(23)	(35)	(18)
Huntington's chorea	1.64 ± 0.09	1.40 ± 0.08[a]	1.52 ± 0.07	1.37 ± 0.16[b]	1.30 ± 0.15[b]	2.75 ± 0.34[b]	2.59 ± 0.18[b]	3.18 ± 0.39	1.72 ± 0.09[c]	4.63 ± 0.35
	(20)	(22)	(20)	(23)	(8)	(9)	(23)	(13)	(30)	(8)
Dialysis encephalopathy	0.96 ± 0.04[b]	0.99 ± 0.06[b]	1.18 ± 0.10[a]	1.91 ± 0.11[b]	2.35 ± 0.20	6.40 ± 0.35	5.53 ± 0.37	3.35 ± 0.30	1.20 ± 0.10[b]	3.86 ± 0.28[c]
	(10)	(10)	(10)	(10)	(10)	(10)	(10)	(9)	(10)	(10)

Note: Values (mean ±SEM) are expressed in μmol/g wet weight. Figures in parentheses indicate number of brains examined.

[a] $p < 0.01$.
[b] $p < 0.001$.
[c] $p < 0.05$.

extensive loss of neurons in the globus pallidus where neuropathological changes are maximal and because of the loss of GABAergic neurons running from or through the globus pallidus to the substantia nigra.[50]

Several other disorders are listed in Table 4 in which the brain contents of GABA are known or suspected to be abnormal. We have found that the content of GABA in both frontal and occipital cortex is significantly reduced in Alzheimer's disease, being as much reduced as it is in these two regions in dialysis encephalopathy. GABA contents have been reported to be low in the nucleus accumbens and in the thalamus in some patients dying with schizophrenia.[51] It has been suggested that tardive dyskinesia, a common neurological complication of long-term use of neuroleptic drugs in schizophrenic patients, may be due to neuroleptic-induced degeneration of the striatopallidal and/or striatonigral GABAergic neurons,[52] and GABA deficiency has been documented in the medial globus pallidus, substantia nigra, and subthalamic nucleus of monkeys treated chronically with fluphenazine or haloperidol.[53] Brains from human patients dying with tardive dyskinesia do not seem to have been examined yet for GABA deficiency in the basal ganglia. CSF concentrations of GABA are markedly reduced in some multiple sclerosis patients who have severe action tremor, and the reports that the action tremor is relieved by GABA-T inhibiting drugs which can increase brain GABA content[54,55] suggest that local deficiencies of GABA may occur in multiple sclerosis.

Table 4 also lists the circumstances in which brain GABA contents may be elevated in humans. In patients dying with idiopathic Parkinson's disease, GABA content is significnatly increased in the putamen,[56,57] presumably as a secondary effect of nigrostriatal neuronal death and decreased release of dopamine in the striatum. Patients with the rare autosomal recessive disorder succinic semialdehyde dehydrogenase deficiency show marked reduction in the activity of this enzyme in lymphocytes isolated from blood, and they accumulate γ-hydroxybutyric acid in blood, urine, and CSF.[58] Although amounts of GABA, the immediate precursor of succinic semialdehyde, have not yet been reported for the CSF and brain of such patients, it would not be surprising if they eventually prove to be elevated.

Finally, drugs which inhibit GABA-T, the first of the two sequential enzymes involved in the degradation of GABA to succinic acid, elevate brain GABA content in experimental animals. Hydrazine, a minor metabolite of isoniazid, is a potent GABA-T inhibitor,[59] and patients treated for Huntington's chorea with isoniazid show a marked rise in CSF GABA concentrations,[60] as do patients treated with γ-vinyl GABA[61] or γ-acetylenic GABA.[62] Examination of the brain of a Huntington's chorea patient who died while under treatment with isoniazid showed partial correction of the expected GABA deficiency in the basal ganglia and overcorrection of GABA contents in other brain regions.[63]

VI. CONCLUSION

Measurements of glutamine, glutamate, and GABA in plasma, CSF, and the brain have contributed important information to our knowledge of the underlying biochemical abnormalities in human brain disorders, and it is likely that careful measurement of these compounds in the brain in heretofore unstudied diseases will lead to further advances. It is of great importance that the analytic methodology be flawless, especially in studies of CSF, where compounds like glutamic acid and GABA which are present in very low concentrations can be artefactually elevated by hydrolysis of precursors. Our ability to increase brain GABA levels with GABA-T inhibiting drugs may lead to useful therapies for presently untreatable neurological disorders. This appears to be the case in Huntington's chorea, where preliminary clinical trials suggest that the relentless neurological deterioration characteristic of this disease is slowed in some patients by long-term isoniazid therapy.[60,64] The same may be true for tardive dyskinesia, where a trial of γ-vinyl GABA produced clinical improvement in most patients,[65] as well as for action tremor in multiple sclerosis.[54,55] Dialysis encephalopathy has not yet been treated with a GABA-T inhibitor, but it might respond favorably.

ACKNOWLEDGMENTS

Work discussed in this chapter from the author's laboratory was supported by the Medical Research Council of Canada and the Huntington Society of Canada. I thank Mrs. Shirley Hansen for her help in preparing this review.

REFERENCES

1. **Perry, T. L., Hansen, S., Wall, R. A., and Gauthier, S. G.,** Human CSF GABA concentrations: revised downward for controls, but not decreased in Huntington's chorea, *J. Neurochem.*, 38, 766, 1982.
2. **Perry, T. L. and Hansen, S.,** Technical pitfalls leading to errors in the quantitation of plasma amino acids, *Clin. Chim. Acta*, 25, 53, 1969.
3. **Perry, T. L., Hansen, S., and Kennedy, J.,** CSF amino acids and plasma — CSF amino acid ratios in adults, *J. Neurochem.*, 24, 587, 1975.
4. **Perry, T. L.,** Cerebral amino acid pools, in *Handbook of Neurochemistry*, Vol. 1, 2nd ed., Lajtha, A., Ed., Plenum Press, New York, 1982, 151.
5. **Minard, F. N. and Mushawar, I. K.,** Synthesis of γ-aminobutyric acid in brain after decapitation, *Life Sci.*, 5, 1409, 1966.
6. **Perry, T. L., Hansen, S., and Gandham, S. S.,** Postmortem changes of amino compounds in human and rat brain, *J. Neurochem.*, 36, 402, 1981.
7. **Marstein, S., Jellum, E., Halpern, B., Eldjarn, L., and Perry, T. L.,** Biochemical studies of erythrocytes in a patient with pyroglutamic acidemia (5-oxoprolinemia), *N. Engl. J. Med.*, 295, 406, 1976.
8. **Perry, T. L., Stedman, D., and Hansen, S.,** A versatile lithium buffer elution system for single column automatic amino acid chromatography, *J. Chromatogr.*, 38, 460, 1968.
9. **Perry, T. L., Berry, K., Hansen, S., Diamond, S., and Mok, C.,** Regional distribution of amino acids in human brain obtained at autopsy, *J. Neurochem.*, 18, 513, 1971.
10. **Perry, T. L., Hansen, S., Berry, K., Mok, C., and Lesk, D.,** Free amino acids and related compounds in biopsies of human brain, *J. Neurochem.*, 18, 521, 1971.
11. **Böhlen, P., Schechter, P. J., van Damme, W., Coquillot, G., Dosch, J.-C., and Koch-Weser, J.,** Automated assay of γ-aminobutyric acid in human cerebrospinal fluid, *Clin. Chem.*, 24, 256, 1978.
12. **Wood, J. H., Hare, T. A., Glaeser, B. S., Ballenger, J. C., and Post, R. M.,** Low cerebrospinal fluid γ-aminobutyric acid content in seizure patients, *Neurology*, 29, 1203, 1979.
13. **Manyam, N. V. B., Katz, L., Hare, T. A., Gerber, J. C., III, and Grossman, M. H.,** Levels of γ-aminobutyric acid in cerebrospinal fluid in various neurological disorders, *Arch. Neurol.*, 37, 352, 1980.
14. **Hare, T. A. and Manyam, N. V. B.,** Rapid and sensitive ion-exchange fluorometric measurement of γ-aminobutyric acid in physiological fluids, *Anal. Biochem.*, 101, 349, 1980.
15. **Perry, T. L., Hansen, S., and Urquhart, N.,** GABA in Huntington's chorea, *Lancet*, 1, 995, 1974.
16. **Armstrong, M. D. and Stave, U.,** A study of plasma free amino acid levels. II. Normal values for children and adults, *Metabolism*, 22, 561, 1973.
17. **Bos, K. D. and Slump, P.,** Determination of glutamine and glutamate in plasma of men and women by ion exchange chromatography, *Clin. Chim. Acta*, 152, 205, 1985.
18. **Walser, M.,** Urea cycle disorders and other hereditary hyperammonemic syndromes, in *The Metabolic Basis of Inherited Disease*, 5th ed., Stanbury, J. B., Wyngaarden, J. B., Fredrickson, D. S., Goldstein, J. L., and Brown, M. S., Eds., McGraw-Hill, New York, 1983, 402.
19. **Kvamme, E.,** Ammonia metabolism in the CNS, *Prog. Neurobiol.*, 20, 109, 1983.
20. **Carpenter, T. O., Levy, H. L., Holthrop, M. E., Shih, V. E., and Anast, C. S.,** Lysinuric protein intolerance presenting as childhood osteoporosis: clinical and skeletal response to citrulline therapy, *N. Engl. J. Med.*, 312, 290, 1985.
21. **Watanabe, A., Takei, N., Higashi, T., Shiota, T., Nakatsukasa, H., Fujiwara, M., Sakata, T., and Nagashima, H.,** Glutamic acid and glutamine levels in serum and cerebrospinal fluid in hepatic encephalopathy, *Biochem. Med.*, 32, 225, 1984.

22. **Perry, T. L., Wirtz, M. L. K., Kennaway, N. G., Hsia, Y. E., Atienza, F. C., and Vemura, H. S.,** Amino acid and enzyme studies of brain and other tissues in an infant with argininosuccinic aciduria, *Clin. Chim. Acta,* 105, 257, 1980.
23. **Perry, T. L., Haworth, J.C., and Robinson, B. H.,** Brain amino acid abnormalities in pyruvate carboxylase deficiency, *J. Inher. Metab. Dis.,* 8, 63, 1985.
24. **Perry, T. L., Hansen, S., and MacLean, J.,** Cerebrospinal fluid and plasma glutamine elevation by anticonvulsant drugs: a potential diagnostic and therapeutic trap, *Clin. Chim. Acta,* 69, 441, 1976.
25. **Marescaux, C., Warter, J. M., Laroye, M., Rumbach, L., Micheletti, G., Koehl, C., Imler, M., and Kurtz, D.,** Le valproate de sodium: une drogue hyperammonemiante: étude chez l'epileptique et chez le volontaire sain, *J. Neurol. Sci.,* 58, 195, 1983.
26. **Batshaw, M. L. and Brusilow, S. W.,** Valproate-induced hyperammonemia, *Ann. Neurol.,* 11, 319, 1982.
27. **Murphy, J. V. and Marquardt, K.,** Asymptomatic hyperammonemia in patients receiving valproic acid, *Arch. Neurol.,* 39, 591, 1982.
28. **Williams, C. A., Tiefenbach, S., and McReynolds, J. W.,** Valproic acid-induced hyperammonemia in mentally retarded adults, *Neurology,* 34, 550, 1984.
29. **Plaitakis, A., Berl, S., and Yahr, M. D.,** Abnormal glutamate metabolism in an adult-onset degenerative neurological disorder, *Science,* 216, 193, 1982.
30. **Plaitakis, A., Berl, S., and Yahr, M. D.,** Neurological disorders associated with deficiency of glutamate dehydrogenase, *Ann. Neurol.,* 15, 144, 1984.
31. **Perry, T. L.,** Four biochemically different types of dominantly inherited olivopontocerebellar atrophy, in *The Olivopontocerebellar Atrophies,* Duvoisin, R. C. and Plaitakis, A., Eds., Raven Press, New York, 1984, 205.
32. **Konigsmark, B. W. and Weiner, L. P.,** The olivopontocerebellar atrophies: a review, *Medicine,* 49, 227, 1970.
33. **Perry, T. L., Kish, S. J., Hinton, D., Hansen, S., Becker, L. E., and Gelfand, E. W.,** Neurochemical abnormalities in a patient with ataxia-telangiectasia, *Neurology,* 34, 187, 1984.
34. **McBride, W. J., Nadi, N. S., Altman, J., and Aprison, M. H.,** Effects of selective doses of X-irradiation on the levels of several amino acids in the cerebellum of the rat, *Neurochem. Res.,* 1, 141, 1973.
35. **Guidotti, A., Biggio, G., and Costa, E.,** 3-Acetylpyridine: a tool to inhibit the tremor and the increase of cGMP content in cerebellar cortex elicited by harmaline, *Brain Res.,* 96, 201, 1975.
36. **Roffler-Tarlov, S. and Sidman, R. L.,** Concentrations of glutamic acid in cerebellar cortex and deep nuclei of normal mice and weaver, staggerer and nervous mutants, *Brain Res.,* 142, 269, 1978.
37. **Perry, T. L. and Hansen, S.,** Amino acid abnormalities in epileptogenic foci, *Neurology,* 31, 872, 1981.
38. **Coyle, J. T., McGeer, E. G., McGeer, P. L., and Schwarcz, R.,** Neostriatal injections: a model for Huntington's chorea, in *Kainic Acid as a Tool in Neurobiology,* McGeer, E. G., Olney, J. W., and McGeer, P. L., Eds., Raven Press, New York, 1978, 139.
39. **Kim, J. S., Kornhuber, H. H., Schmid-Burgk, W., and Holzmüller, B.,** Low cerebrospinal fluid glutamine in schizophrenic patients and a new hypothesis on schizophrenia, *Neurosci. Lett.,* 20, 379, 1980.
40. **Kornhuber, H. H.,** Chemistry, physiology and neuropsychology of schizophrenia: towards an earlier diagnosis of schizophrenia. I, *Arch. Psychiatr. Nervenkr.,* 233, 415, 1983.
41. **Perry, T. L.,** Normal cerebrospinal fluid and brain glutamate levels in schizophrenia do not support the hypothesis of glutamatergic neuronal dysfunction, *Neurosci. Lett.,* 28, 81, 1982.
42. **Perry, T. L., Hansen, S., and Kloster, M.,** Huntington's chorea: deficiency of γ-aminobutyric acid in brain, *N. Engl. J. Med.,* 288, 337, 1973.
43. **Glaeser, B. S., Vogel, W. H., Oleweiler, D. B., and Hare, T. A.,** GABA levels in cerebrospinal fluid of patients with Huntington's chorea: a preliminary report, *Biochem. Med.,* 12, 380, 1975.
44. **Enna, S. J., Stern, L. Z., Wastek, G. J., and Yamamura, H. I.,** Cerebrospinal fluid γ-aminobutyric acid variations in neurological disorders, *Arch. Neurol.,* 34, 684, 1977.
45. **Sweeney, V. P., Perry, T. L., Price, J. D. E., Reeve, C. E., Godolphin, W. J., and Kish, S. J.,** Brain γ-aminobutyric acid deficiency in dialysis encephalopathy, *Neurology,* 35, 180, 1985.
46. **Perry, T. L., Yong, V. W., Kish, S. J., Ito, M., Foulks, J. G., Godolphin, W. J., and Sweeney, V. P.,** Neurochemical abnormalities in brains of renal failure patients treated by repeated hemodialysis, *J. Neurochem.,* 45, 1043, 1985.
47. **Obata, K.,** Association of GABA with cerebellar Purkinje cells: single cell analysis, in *GABA in Nervous System Function,* Roberts, E., Chase, T., and Tower, D. B., Eds., Raven Press, New York, 1976, 217.
48. **Storm-Mathisen, J.,** Distribution of the components of the GABA system in neuronal tissue: cerebellum and hippocampus — effects of axotomy, in *GABA in Nervous System Function,* Roberts, E., Chase, T., and Tower, D. B., Eds., Raven Press, New York, 1976, 149.
49. **Roberts, E.,** New directions in GABA research. I. Immunocytochemical studies of GABA neurons, in *GABA-Neurotransmitters: Pharmacochemical, Biochemical, and Pharmacological Aspects,* Krogsgaard-Larsen, P., Scheel-Krüger, J., and Kofod, H., Eds., Academic Press, New York, 1979, 28.

50. **Perry, T. L., Norman, M. G., Yong, V. W., Whiting, S., Crichton, J. V., Hansen, S., and Kish, S. J.,** Hallervorden-Spatz disease: cysteine accumulation and cysteine dioxygenase deficiency in the globus pallidus, *Ann. Neurol.*, 18, 482, 1985.
51. **Perry, T. L., Kish, S. J., Buchanan, J., and Hansen, S.,** γ-Aminobutyric-acid deficiency in brain of schizophrenic patients, *Lancet*, 1, 237, 1979.
52. **Fibiger, H. C. and Lloyd, K. G.,** Neurobiological substrates of tardive dyskinesia: the GABA hypothesis, *TINS*, 7, 462, 1984.
53. **Gunne, L.-M., Häggstrom, J.-E., and Sjöquist, B.,** Association with persistent neuroleptic-induced dyskinesia of regional changes in brain GABA synthesis, *Nature (London)*, 309, 347, 1984.
54. **Sabra, A. F., Hallett, M., Sudarsky, L., and Mullally, W.,** Treatment of action tremor in multiple sclerosis with isoniazid, *Neurology*, 32, 912, 1982.
55. **Hallett, M., Lindsay, J. W., Adelstein, B. D., and Riley, P. O.,** Controlled trial of isoniazid therapy for severe postural cerebellar tremor in multiple sclerosis, *Neurology*, 35, 1374, 1985.
56. **Perry, T. L., Javoy-Agid, F., Agid, Y., and Fibiger, H. C.,** Striatal GABAergic neuronal activity is not reduced in Parkinson's disease, *J. Neurochem.*, 40, 1120, 1983.
57. **Kish, S. J., Rajput, A., Gilbert, J., Rozdilsky, B., Chang, L.-J., Shannak, K., and Hornykiewicz, O.,** Elevated gamma-aminobutyric acid level in striatal but not extrastriatal brain regions in Parkinson's disease: correlation with striatal dopamine loss, *Ann. Neurol.*, 20, 26, 1986.
58. **Gibson, K. M., Sweetman, L., Nyhan, W. L., Jakobs, C., Rating, D., Siemes, H., and Hanefeld, F.,** Succinic semialdehyde dehydrogenase deficiency: an inborn error of gamma-aminobutyric acid metabolism, *Clin. Chim. Acta*, 133, 33, 1983.
59. **Perry, T. L., Kish, S. J., Hansen, S., Wright, J. M., Wall, R. A., Dunn, W. L., and Bellward, G. D.,** Elevation of brain GABA content by chronic low-dosage administration of hydrazine, a metabolite of isoniazid, *J. Neurochem.*, 37, 32, 1981.
60. **Perry, T. L., Wright, J. M., Hansen, S., Baker Thomas, S. M., Allan, B. M., Baird, P. A., and Diewold, P. A.,** A double-blind clinical trial of isoniazid in Huntington disease, *Neurology*, 32, 354, 1982.
61. **Grove, J., Schechter, P. J., Tell, G., Koch-Weser, J., Sjoerdsma, A., Warter, J. M., Marescaux, C., and Rumbach, L.,** Increased gamma-aminobutyric acid (GABA), homocarnosine and β-alanine in cerebrospinal fluid of patients treated with γ-vinyl GABA (4-amino-hex-5-enoic acid), *Life Sci.*, 28, 2431, 1981.
62. **Tell, G., Böhlen, P., Schechter, P. J., Koch-Weser, J., Agid, Y., Bonnet, A. M., Coquillat, G., Chazot, G., and Fischer, C.,** Treatment of Huntington disease with γ-acetylenic GABA, an irreversible inhibitor of GABA-transaminase: increased CSF GABA and homocarnosine without clinical amelioration, *Neurology*, 31, 207, 1981.
63. **Perry, T. L., Wall, R. A., and Hansen, S.,** Brain amino compounds in a Huntington's disease patient on isoniazid therapy, *Neurology*, 35, 755, 1985.
64. **Stober, T., Schimrigk, K., Holzer, G., and Ziegler, B.,** Quantitative evaluation of functional capacity during isoniazid therapy in Huntington's disease, *J. Neurol.*, 229, 237, 1983.
65. **Stahl, S. M., Thornton, J. E., Simpson, M. L., Berger, P. A., and Napoliello, M. J.,** Gamma-vinyl-GABA treatment of tardive dyskinesia and other movement disorders, *Biol. Psychiatr.*, 20, 888, 1985.

Chapter 22

PATHOLOGY OF GLUTAMATE DEHYDROGENASE

Andreas Plaitakis and Soll Berl

TABLE OF CONTENTS

I. INTRODUCTION

Considerable amounts of glutamate dehydrogenase (GDH) activity are known to be present in the mammalian brain.[1-4] Developmental studies have shown that GDH activity increases in the brain postnatally; these changes parallel the increase in brain glutamate content and nerve terminal growth.[1-3] Glutamate is thought to serve as the main excitatory transmitter in mammalian nervous tissue.[5] However, the compound is known to have neuroexcitotoxic potentials[6] and it may mediate neuronal damage in response to various insults.[7,8] Furthermore, because the local administration of glutamate and particularly its potent analogues can produce in the brain morphological and biochemical alterations that are similar to those found in patients with neurodegenerative disorders,[9-11] the possibility has been raised that neuroexcitotoxic mechanisms may be involved in the pathogenesis of these human diseases. Several enzymes related to glutamate, aspartate, and GABA metabolism such as glutamate decarboxylase (GAD) and aspartate aminotransferase (AAT) have been studied in neural tissues of patients dying from degenerative diseases;[12-14] however, these studies have not provided evidence that abnormalities of these enzymes are causally related to the neurodegenerative process(es) in these disorders.

On the other hand, clues implicating a possible role of the NADP(H)-dependent oxidoreductases in mechanisms of selective neuronal degeneration were provided by studies related to the neurotoxic effects of the nicotinamide antagonist 3-acetylpyridine (3-AP). The systemic administration of the compound to rats has resulted in selective destruction of neurons located in the inferior olives, lower cranial nerve nuclei, pons, and substantia nigra.[15] Because the histopathology produced by the toxic agent is similar to that found in patients with olivopontocerebellar degeneration (OPCA),[16,17] it was thought that the selective neurotoxicity of 3-AP could provide clues for elucidating the biochemical pathology of OPCA,[18] a situation analogous to the use of 1-methyl-4-phenyl-1,2,5,6-tetrahydropyridine (MPTP) toxicity as a model for Parkinson's disease.[19]

Previous studies had shown that 3-AP, acting as a nicotinamide antagonist, is incorporated into the NADP(H) system forming substantial amounts of 3-APADP(H) in the brain of experimental animals.[20] Because 3-APADP(H) has been shown to inhibit in vitro NADP(H)-dependent oxidoreductases,[20] these enzymes were evaluated in cultured skin fibroblasts from a patient with OPCA and results revealed a selective reduction in the activity of GDH.[17] Since this original observation, GDH activities have been measured in leukocytes or platelets obtained from peripheral blood of patients with degenerative neurological disorders of various types. A partial deficiency of GDH activity (reduction by 40 to 60%) measured in whole tissue homogenates has been found in patients with clinically diagnosed OPCA.[21-27]

II. NEUROLOGICAL DISORDERS ASSOCIATED WITH GDH DEFICIENCY

A. General

The OPCAs are adult-onset neurological disorders which may occur either sporadically or be genetically transmitted (dominantly or recessively).[28] They are characterized clinically by cerebellar ataxia which usually occurs in association with other neurological deficits such as parkinsonism, oculomotor disturbances, and/or motor neuron signs and pathologically by degeneration of the inferior olives, gray matter of the pons, cerebellar cortex, lower cranial nerve nuclei, substantia nigra, and spinal cord. Because of these characteristics, the OPCAs occupy a central position among the various neurodegenerative diseases linking the spinocerebellar degenerations with the extrapyramidal diseases. Although the term OPCA was first used by Dejerine and Thomas in 1900 to refer to sporadic cases of the disease, the majority of the clinicopathological studies published during the past 9 decades have been on families with a dominantly inherited form of the disease; the recessively inherited form might have been underrepresented in world literature on this subject.[16]

Several large kindreds with dominant OPCA have been studied in detail.[28-30] In these, the onset of the disease has usually occurred in the 2nd and 3rd decade of life with affected members found in successive generations. The mode of genetic transmission has been shown to be dominant with complete penetrance. Here, it is referred to as *dominant OPCA*.

OPCA may also occur in association with clinical signs of progressive autonomic dysfunction,[28,31] with orthostatic hypotension being a consistent feature. This disorder is distinct from the other forms of OPCA and usually occurs sporadically in late adult life (5th-6th decades). Here, it is referred to as *OPCA with orthostatic hypotension*.

Studies on the specificity of GDH deficiency centered around the above nosological concepts of OPCA and its relationship to other forms of spinocerebellar degeneration and basal ganglia disorders. Early in the course of these studies, it was found that the enzyme defect was not present in several large kindreds affected by early onset dominant OPCA with complete penetrance.[21,23] Also, patients with sporadic, late-onset OPCA associated with orthostatic hypotension were found to have normal GDH activity in their leukocytes.[23]

In addition, other forms of spinocerebellar degeneration such as Friedreich's ataxia (FA), cerebello-olivary atrophy (COA), familial spastic paraplegia (FSP), and Charcot-Marie-Tooth disease (CMTD) had normal GDH activity. The same was also true for several basal ganglia disorders such as Parkinson's disease (PD), Huntington's chorea, and progressive supranuclear palsy (PSP).[23] As described below, GDH deficiency was found to be associated with certain late-onset forms of OPCA which were not associated with orthostatic hypotension[17,21-24] (Figure 1).

B. Clinical Phenotypes

The clinical characteristics of patients with GDH deficiency have been described in detail in at least three studies.[17,23,26] The majority of the reported cases experienced their first symptoms during the 5th or 6th decade of their lives. Gait disturbances such as unsteadiness on walking or lack of balance (with a particular difficulty in turning or going downstairs) and sometimes slowness of gait or "hesitation in step" have been the most common presenting symptoms of the disease. However, a few patients experienced disturbances in oculomotor function causing diplopia or blurriness of vision, whereas others noted difficulty with their speech (dysarthria) early in the course of their illness. As the disease slowly evolved, increasing gait disturbances (resulting in repeated falls), difficulty with fine hand coordination, slowness of movements, and deteriorating speech occurred. Neurological examination revealed cerebellar ataxia in most cases in association with other neurological deficits such as dysarthria, oculomotor disturbances, parkinsonism, corticospinal signs, amyotrophy, and peripheral neuropathy occurring in various combinations. A considerable variation in the clinical presentation has been noted among patients with the enzymatic deficiency.[23] Thus, some patients showed a predominant cerebellar ataxia with or without peripheral neuropathy, whereas others showed signs of parkinsonism (atypical Parkinson's disease) or motor neuronal degeneration (secondary ALS). These symptoms and signs of the disease were thought to result from atrophic changes involving the cerebellar system and its interconnective brain stem structures (inferior olives and pontine nuclei) as well as the basal ganglia and the spinal cord. Post-mortem examination in patients dying from this disease have indeed confirmed this concept[37] (see below).

C. Genetic Considerations

Although the index case of GDH deficiency was a sporadic occurrence,[17] subsequent studies indicated the familiar incidence of the disorder.[21-27] In about half of the patients we have studied,[23] the disease has affected siblings of both sexes, but no parents or offspring.

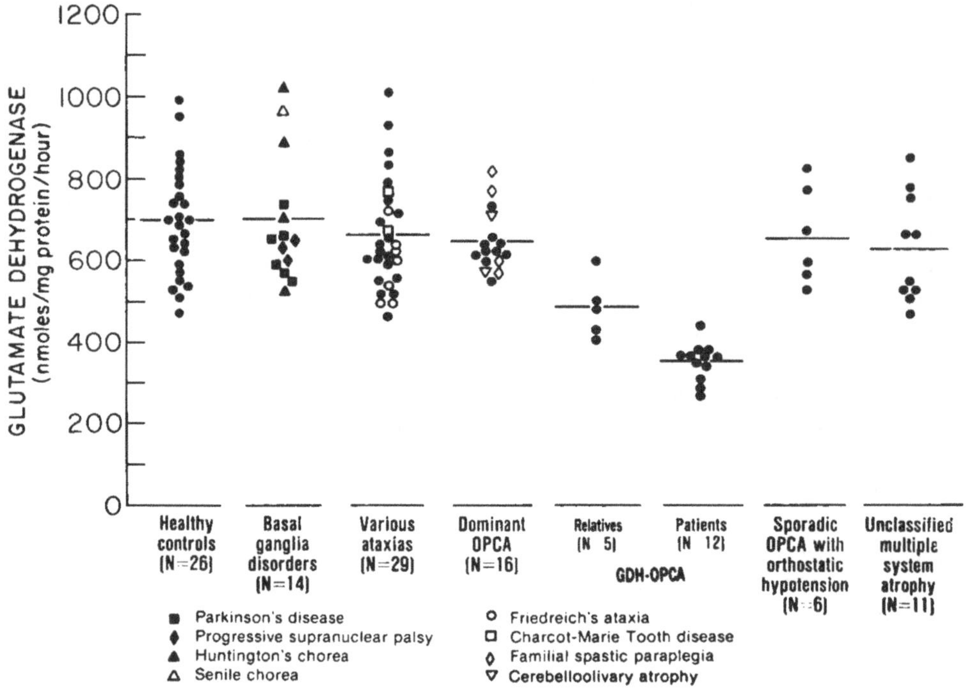

FIGURE 1. GDH activity in leukocytes from healthy controls and patients with degenerative neurological disorders of various types. Enzyme activity was measured fluorometrically in whole homogenates in the presence of 1 m*M* ADP (no detergent was used).[24] Values are the average of several determinations. At least three leukocyte homogenates were studied from each patient with GDH-OPCA. Under various ataxias are included patients with nondominant form of cerebellar and spinocerebellar disorders other than OPCA. Under dominant OPCA are included ten patients (closed symbols) affected by early adult-onset dominant OPCA with complete penetrance (five kindreds).[28-30] Under GDH-OPCA (patients) are included patients with late adult-onset OPCA associated with consistently decreased GDH activity in leukocytes. Seven of these patients were familiar (possibly recessively inherited) and five were sporadic occurrences. They were grouped together because of the low GDH activity found in their leukocytes and the similar nosological features they represented. Under relatives are included two asymptomatic parents and three offspring of patients with OPCA associated with GDH deficiency. Under unclassified multiple system atrophy are included patients with late adult-onset sporadic extrapyramidal syndromes associated with other neurologic deficits that couldn't be classified with certainty. One patient came to post-mortem examination which revealed a variant of nigrostriatal degeneration.[23]

We have considered these genetic data to suggest the recessive inheritance of the enzymatic defect,[21-23] although we could not rule out the possibility of a dominant transmission with incomplete penetrance.[21] A limited number of parents and offspring of affected patients have been studied.[23] Intermediate leukocytic GDH values in five of these individuals were found and considered as compatible with the excessive inheritance state[23] (Figure 1). On the other hand, Duvoisin et al.[26] have presented evidence supporting the possibility of a dominant transmission of the enzymatic defect. However, when these authors studied larger numbers of family members, they also raised the possibility of recessive transmission of the disorder.[27] There are two main difficulties with doing genetic studies in patients with GDH deficiency. The first is due to the late onset of the disease and its variable course and expressivity which precludes detailed studies of families spanning several generations, and the second is due to the magnitude of the enzymatic defect which makes identification of the heterozygous state problematic.

D. Nature of the Enzymatic Defect

Although it is yet uncertain as to whether or not partial enzymatic defects are capable of

Table 1
EFFECT OF ADP AND TRITON® X-100 ON GDH ACTIVITY OF WHOLE LEUKOCYTIC HOMOGENATES FROM CONTROLS

	GDH in leukocytes[a] (nmol/mg protein/hr; mean +SD)		
GDH	+ADP (1 m*M*)	+Triton® X-100 (0.025%)	+ADP (1 m*M*), +Triton® X-100 (0.025%)
384 ± 102 (N = 5)	702 ± 213 (N = 5) $p<0.01$	362 ± 100 (N = 5)	1642 ± 210 (N = 10) $p<0.001$

[a] N = Number of leukocyte homogenates studied.

From Plaitakis, A., *Ital. J. Neural Sci. Suppl.*, 4, 65, 1984. With permission.

causing late-onset degenerations, the magnitude of decrease in GDH activity (by about 50%) is not entirely consistent with a recessively inherited primary enzymatic deficiency. Kinetic studies, carried out with whole homogenates of cultured skin fibroblasts of the original OPCA case, failed to show a significant alteration in the K_ms for the substrates of the enzyme, α-ketoglutarate (2-oxoglutarate), NADH, and ammonium.[17]

Early studies with leukocyte and fibroblast homogenates were performed without the use of detergents.[17] Most workers in the field have not used detergents for estimation of GDH activity in tissue homogenates because no appreciable increase in GDH activity can be obtained by using these agents which may be inhibitory to GDH.[31a] Since the prevailing belief has been that GDH is a mitochondrial matrix enzyme that is easily released in a soluble form by tissue homogenization,[31a-34] its estimation was not expected to be greatly influenced by the presence of a detergent during assay.

In accord with this concept, our initial studies[17] showed that Triton® X-100 (1.6%) used in the absence of a GDH activator did not increase the enzyme activity of homogenates, but it was, in fact, inhibitory to GDH. However, further studies[23,24] indicated that ADP (1 m*M*) or L-leucine (5 to 10 m*M*) had a much greater stimulatory effect on GDH activity of whole homogenates of leukocytes in the presence of low concentrations (0.025 to 0.05%) of Triton® X-100 or Lubrol® (by 300 to 450%) than in the absence of these detergents (by 75 to 120%) (Table 1). At such low concentrations, Triton® X-100 and Lubrol® did not alter the baseline activity (measured without the use of ADP or L-leucine) of the homogenates (Table 1). When whole leukocyte homogenates were fractionated into soluble (supernatant) and particulate (pellet) fractions by centrifugation, it was found that the detergents had no significant effect on the activation of the soluble GDH activity by ADP or L-leucine; however, these activators produced a disproportionate stimulation of the particulate GDH activity that was even greater than that observed with whole homogenates in the presence of the detergents.

On the basis of these experiments, it was concluded that soluble and particulate GDH activities are present in tissue homogenates[23] and that the latter is markedly underestimated under routine assay conditions unless both a detergent and an activator are present in the reaction mixture during assay.

When GDH assays were performed in cultured skin fibroblasts of five OPCA patients (all of whom had decreased enzyme activities in their leukocytes), it was found that low concentrations of Triton® X-100 (0.025%), in the presence of 1 m*M* ADP, stimulated markedly the enzymatic activity of controls, but had little effect on the GDH activity of cell homog-

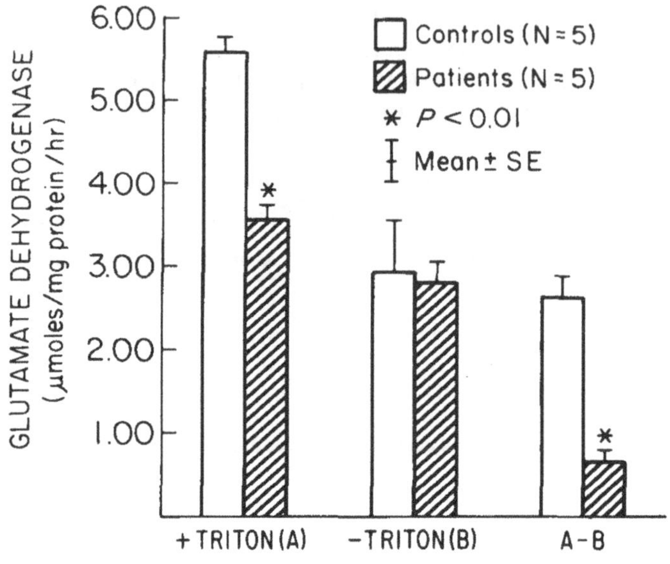

FIGURE 2. GDH activity in whole homogenates of cultured skin fibroblasts from patients and controls. Skin was obtained by biopsy after informed consent was secured. Primary cultures of skin fibroblasts were established and passed as previously described.[17] Fibroblast pellets were disrupted by three to four cycles of freeze-thaw and whole homogenates were prepared.[17] GDH activity was determined as previously described[17] (with the addition of 1 m*M* ADP) to the reaction mixture in the presence (+Triton®) or absence (−Triton®) of Triton® X-100 (0.025% final concentration). (From Plaitakis, A., *Ital. J. Neural Sci. Suppl.*, 4, 65, 1984. With permission.)

enates from patients.[24] Estimation of the "Triton®-activated" GDH activity revealed that the values of the patients were about 20% of those found in controls, whereas total activity values (obtained with the addition of Triton® X-100 and ADP to the reaction mixture) were 60% of those of controls (Figure 2). These results were considered suggestive of a selective defect in the particulate GDH activity of the fibroblasts of the patients.

This possibility was tested with the use of leukocytes which were more readily available than the cultured cells. Whole leukocytic homogenates were subject to a low-speed centrifugation (480 xg for 10 min) in order to remove nuclei and large fragments of cells. The resulting low-speed supernatant was further fractionated into high-speed supernatant and high-speed pellet (100,000 xg for 1 hr) containing the soluble and particulate (membrane-associated) GDH activity. Determination of these fractions in OPCA patients revealed that the decrease in activity was limited to the particulate GDH component.[23]

Since these data suggested that the two GDH fractions may be under different genetic control, the possibility that the two GDH fractions represent distinct isoenzymes was further investigated. Studies on partially purified leukocytic fractions revealed the two activities to differ significantly in thermal stability characteristics.[23] Thus, when activities of the two enzymes from control leukocytes were incubated at 47.5°C in 50 m*M* Tris-HCL, pH 7.40, buffer, the half-life of the soluble fractions was 10 hr, whereas that of the particulate fraction was 24 min. Increasing the pH of the buffer to 8.00 and 8.50 increased the rate of heat inactivation of the particulate enzyme to 13 and 4 min, respectively, and increased the soluble enzyme to 4 and 1 hr, respectively.[35] Lower, but proportionately similar, rates of heat inactivation were obtained by incubating the two GDH fractions at 45°C in Tris HCl 50 m*M* buffer of various pHs (Figure 3). The presence of L-leucine (10 m*M*), ADP (1 m*M*),

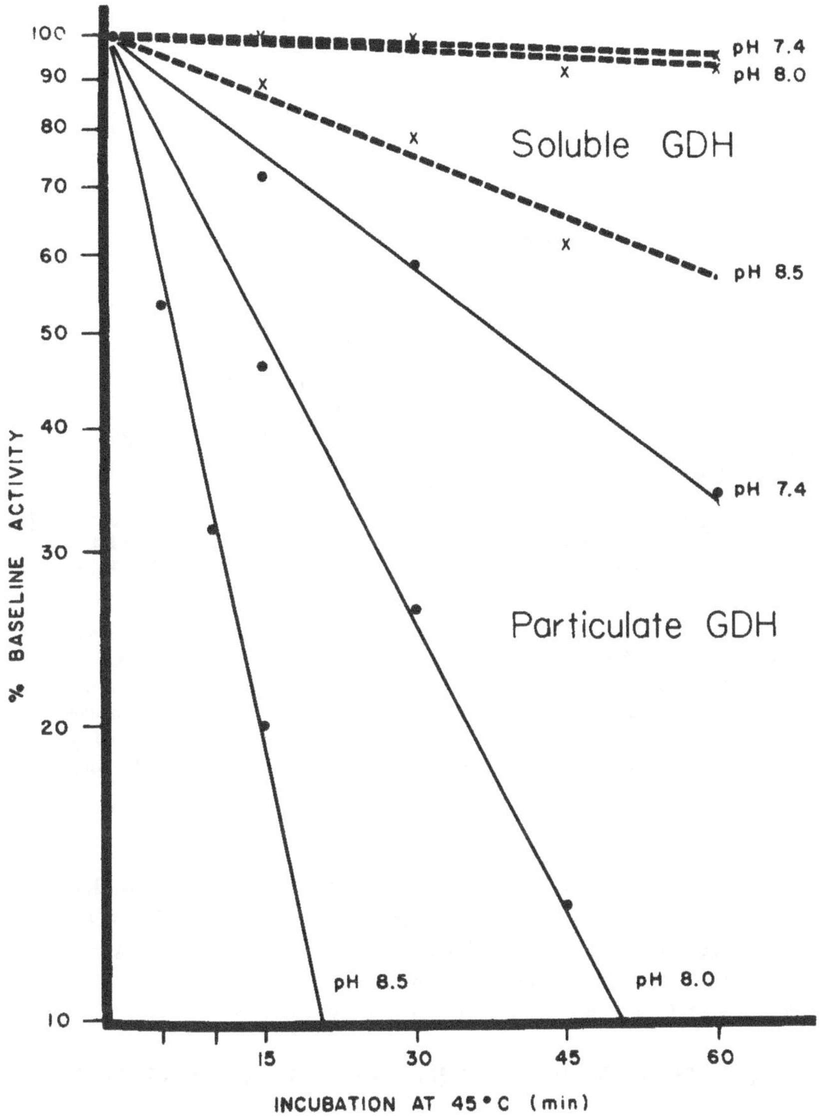

FIGURE 3. Heat inactivation of soluble and particulate leukocytic GDH components. Leukocyte pellets from controls were subjected to three to four cycles of freeze-thaw and then homogenized in 50 m*M* Tris HCl buffer of various pHs. Tissue fractionation was carried out as previously described.[23] Bovine serum albumin was added to each fraction to 2 mg/mℓ. The fractions were incubated with agitation in capped Falcon® plastic 12- × 75-mm tubes at 45°C. The pH values are those of the incubation buffer (50 m*M* Tris HCl). At the time points shown in the figure, portions of each preparation were removed and assayed for GDH activity. ADP (1 m*M*) and Triton® X-100) (0.025%) were added to the reaction mixture at the time of assay.

NaCl (154 m*M*), KCl (154 m*M*), or ethylene glycol (20 to 40%) protected both fractions from heat inactivation.

In view of the above findings, heat-stable and heat-labile forms of GDH were measured in whole homogenates of leukocytes from patients and controls by incubating them at 47.5°C in 50 m*M* Tris HCl, pH 7.40, for 60 min or in 50 m*M* Tris HCl, pH 8.0, for 30 min. Under these conditions, there was inactivation of most of the particulate GDH with very little

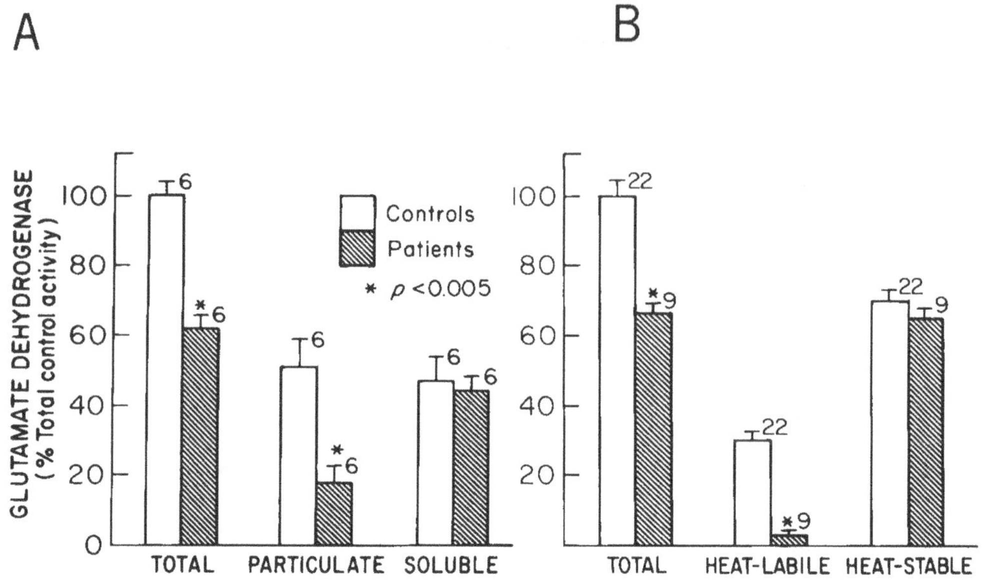

FIGURE 4. (A) GDH activity in whole homogenates (total) of leukocytes and soluble and particulate fractions. Tissue fractionation and GDH assays were carried out as previously described.[23] (B) Activities of heat-labile and heat-stable GDH components in leukocytes from patients and controls. Total GDH activity was measured in whole homogenates in the presence of Triton® X-100 (0.025%) and 1 m*M* ADP, both added to the reaction mixture at the time of assay. The heat-labile and heat-stable GDH activities were determined by heat inactivation at 47.5°C for 1 hr as previously described.[23] All values are expressed as mean ± SD. Numbers indicate the number of individuals studied.[23]

inactivation of the soluble GDH, thus allowing us to differentiate the two enzyme forms. Results revealed[23] that patients with late-onset form OPCA had a complete deficiency of the heat-labile GDH (8% of control), but normal heat-stable enzyme levels (Figure 4). These results further corroborate the data on leukocyte fractionation studies showing a selective defect in the particulate (heat-labile) GDH in the OPCA patients and further suggest that a mutation in one GDH "isoenzyme" may underly the disease.

III. GLUTAMATE METABOLISM IN PATIENTS WITH GDH DEFICIENCY

Since deficiency of GDH occurs in extraneural tissues (leukocytes and cultured skin fibroblasts) of patients with a particular form of OPCA, the possibility was explored as to whether these patients showed abnormalities in the systemic metabolism of the substrates of the enzyme, glutamate, α-ketoglutarate, and ammonia.[36] Measurement of these compounds in the fasting plasma of GDH-deficient patients revealed that the levels of glutamate were significantly increased (by about 150%, $p < 0.0005$), whereas those of α-ketoglutarate were significantly decreased (by 25%, $p < 0.005$); the levels of ammonia were not altered. Besides the fasting glutamic acidemia (Figure 5), these patients also showed significantly greater increases in plasma glutamate levels than controls following ingestion of a standard meal containing 1 g of protein per kilogram of body weight.[36] Protein is known to contain 11% glutamic acid. In addition, decreased clearance of an orally administered glutamate load (60 mg of monosodium glutamate, MSG, per kilogram of body weight after overnight fasting) has been shown in these patients.[18,37] Following MSG intake, an excessive increase in plasma glutamate levels occurred (Figure 6) associated with proportional increases in the plasma aspartate levels.

Exogenously administered glutamate is expected to be primarily metabolized by the liver

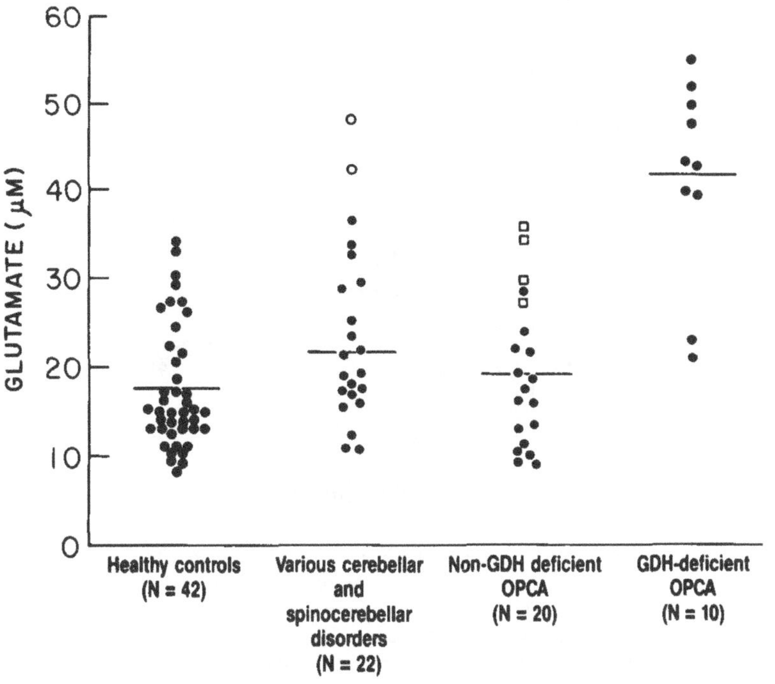

FIGURE 5. Fasting plasma levels of glutamate in patients with GDH deficiency and controls. Lines represent the mean value for each group. N = number of individuals studied. The various cerebellar and spinocerebellar disorders include non-GDH-deficient entities such as Friedreich's ataxia (N = 3), familial spastic paraplegia (N = 5), cerebello-olivary atrophy (N = 2), and miscellaneous syndromes (N = 10). Open circles represent values from patients with ataxic encephalopathy associated with hyperpyruvemia. The non-GDH-deficient OPCA includes 15 patients with dominant OPCA with complete penetrance (from 6 families) and 5 patients with sporadic OPCA associated with orthostatic hypotension. Four patients of the former group (open squares) had a form of dominant OPCA that is associated with slowed saccadic eye movements. (From Plaitakis, A., *Ital. J. Neural Sci. Suppl.*, 4, 65, 1984. With permission.)

and other extraneural tissues such as kidneys, gut, etc., since the compound cannot penetrate the blood-brain barrier. At least two pathways of glutamate metabolism by the mitochondria, coupled to two separate mechanisms of entry into these organelles, have been suggested.[38,39] The first involves entrance via the glutamate-aspartate translocase with resultant obligatory transamination to aspartate (by the AAT) and exit of aspartate. The second mechanism involves entrance of glutamate via the glutamate-OH translocase with resultant oxidation of the compound (by GDH).

Since the plasma aspartate levels increased significantly greater in the GDH-deficient patients after MSG intake as compared to controls and since these changes were proportional to glutamate increases, it may be concluded that the transamination pathway was not affected in these individuals (Figure 7). Consequently, the present data may be reasonably explained by a systemic defect in the oxidative deamination of glutamate to α-ketoglutarate. These results, thus, provide supportive evidence that malfunction of GDH occurs in the tissues of the GDH-deficient patients.

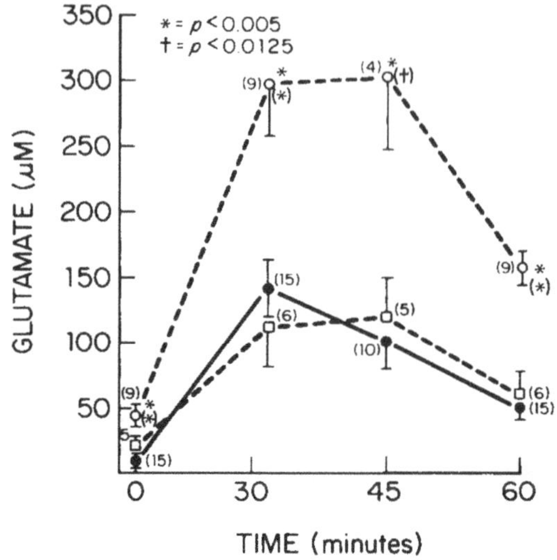

FIGURE 6. Glutamate loading tests in patients with GDH deficiency (open circles), healthy controls (closed circles), and diseased controls (open squares). The latter are patients with sporadic OPCA associated with orthostatic hypotension. Patients and controls received orally 60 mg/kg body weight of MSG after fasting overnight. Blood samples were drawn at the time points indicated and processed and analyzed as previously described.[36] Number in parenthesis indicates the number of individuals studied.

IV. PATHOGENESIS OF THE LOCALIZED BRAIN ATROPHIC CHANGES IN OPCA

A. General

The OPCAs encompass a number of heterogenous neurodegenerative disorders that share common pathologic features, such as atrophy of the cerebellar cortex, inferior olives, olivocerebellar fibers, basis pontis (pontine nuclei), and transverse pontine fibers. In some forms of the disease, the substantia nigra, basal ganglia, cerebral cortex, and spinal cord may also be involved.[16] The pathogenesis of these localized atrophic changes, as with those occurring in other major neurodegenerative disorders with system atrophy, remains unknown. The detection of GDH deficiency in a particular form of this disorder has, however, provided clues toward elucidating the molecular and cellular events that lead to the neurodegenerative process as described below.

B. Evidence from Neuroimaging Studies in OPCA Patients

Recently, high-resolution CT scans of the heads of patients with OPCA of five different types were studied in detail.[40] Results revealed that GDH-deficient OPCA shows a characteristic pattern of brain abnormalities that is distinct from the other forms of the disease that are not associated with the enzymatic deficiency. These included atrophy of the brain stem and cerebellum, enlargement of the third ventricle, widening of the Sylvian fissures and intercaudate distance, and frontal convolutional atrophy. Most of these morphological changes could be attributed to degeneration of structures that receive putative glutamatergic innervation or have been shown experimentally to be sensitive to glutamate toxicity (Figure 8). As already mentioned, the neuroexcitotoxic potentials of glutamate have been well established and alterations in glutamate metabolism have been shown to occur in patients

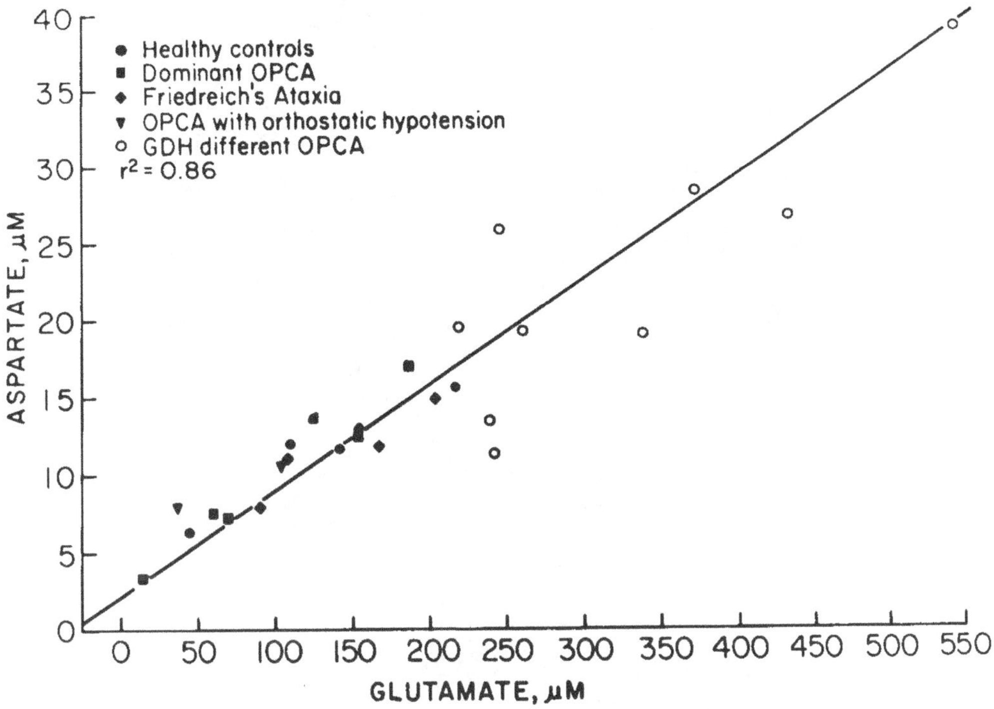

FIGURE 7. Plasma glutamate levels plotted against aspartate levels at 30 min after oral loading with MSG (60 mg/kg body weight). Closed symbols represent values from healthy and diseased controls (dominant OPCA, FA, and OPCA with orthostatic hypotension). Open symbols represent values of GDH-deficient patients. (From Plaitakis, A. and Berl, S., *J. Neural Transm. Supp.*, 13, 67, 1983. With permission.)

with GDH deficiency. A selective defect in the membrane-associated form of GDH has been found in extraneural as well as neural tissues (see below) of such patients. Additional experimental data[41] have shown that in the brain this membrane-bound component of the enzyme is associated with synaptic membranes and may, therefore, be involved with the metabolism of the "synaptic pool" of glutamate.[42] Consequently, these data taken together strongly suggest that alterations in glutamate metabolism may occur at the glutamatergic nerve terminals in the nervous tissue of OPCA patients with resulting degeneration of postsynaptic neurons in accord with the excitotoxic theory.[6-11]

C. Evidence from Post-Mortem Studies

Recently, a 72-year-old patient affected by GDH-deficient OPCA came to post-mortem examination.[37] The patient was followed carefully for about 4 years. Detailed studies on GDH activities in leukocytes and cultured skin fibroblasts and glutamate loading tests had been performed on this patient.

The brain was obtained within 3 hr after death and was divided by sagittal dissection in two halves: one was fixed with formaline for morphological studies and the other was stored at −80°C for biochemical studies.

Pathological examination revealed atrophy of the pons and the cerebellum. Microscopically, there was some atrophy of the cerebellar folia associated with an almost complete disappearance of the Purkinje cells with relative preservation of the granule cell layer. The inferior olives, the pontine nuclei, and the transverse pontine fibers were markedly atrophic. A substantial loss of the pigmented neurons was present in the substantia nigra. In the basal

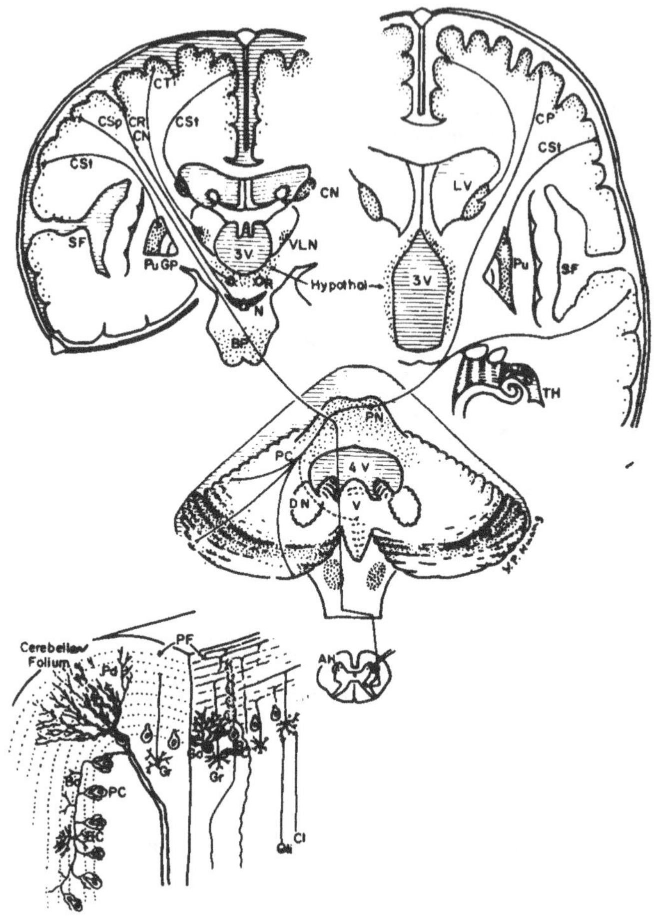

FIGURE 8. Schematic diagrams showing putative glutamatergic corticofugal fibers in relation to morphological changes of the brain in CT patients with GDH deficiency. Coarsely dotted are areas of the brain directly innervated by the putative glutamatergic fiber systems[43] which include the corticopontine tract (CP), corticorubral fibers (CR), corticospinal tract (CSp), corticostriate fibers (CSt), and corticothalamic (CT) fibers. Atrophy of the basis pontis, cerebellum, and frontal lobes as well as enlargement of the third (3v) and fourth (4v) ventricles, widening of the Sylvian fissures (SF), and increase in intercaudate distance are shown schematically. This topography of brain lesions appears to be consistent with their "excitotoxic pathogenesis". According to this process, deficiency in GDH may lead to accumulation of glutamate in neurotoxic amounts at the glutamatergic nerve terminals with resultant degeneration of the postsynaptic neurons.[36] This could account for the degeneration of the pontine nuclei (PN) which are innervated by the corticopontine tract (CP) and give rise to the pontocerebellar fibers (PC). Similar considerations may also apply to the degeneration of Purkinje cells of the cerebellar cortex which receive innervation by the putative glutamatergic granule cells.[44] Atrophy of the caudate nucleus (CN) and putamen (Pu) — which are innervated by the putative glutamatergic corticostriate tract (Cst)[45] — may be responsible for the increased intercaudate distance and Sylvian fissures, respectively, as shown by CT. Involvement of the thalamus, which receives the corticothalamic (CT) and the hypothalamus appears to be the anatomical basis for the enlargement of the anterior third ventricle. Selective damage to the hypothalamus with resultant enlargement of the third ventricle has been produced experimentally by the systemic administration of MSG to premature animals.[6]

ganglia, the most conspicuous change was the severe atrophy of the posteriolateral putamen with necrosis of neurons and gliosis present locally.[37]

Thus, post-mortem examination of the brain in a GDH-deficient patient confirms earlier conclusions, based on the CT studies of a rather large number of patients with this condition, suggestive of a selective involvement of neuronal systems that receive putative glutamatergic innervation. These include the Purkinje cells of the cerebellar cortex that receive input from the granule cells (putative glutamatergic); the latter being presynaptic cells that were relatively spared. The pontine nuclei and the putamen which are known to be innervated by the putative glutamatergic corticopontine and corticostriatal tracks, respectively, were also selective targets of the neurodegenerative process.

D. GDH Deficiency in the Brain of an OPCA Patient

Cerebellar cortical tissue from the above patient and five controls were studied.[37] GDH activity in whole tissue homogenates was partially deficient in the patient (60% of control value). Similar to the results obtained with the use of cultured skin fibroblasts, Triton® X-100 (0.05%) in the presence of 1 m*M* ADP produced little activation of the GDH activity of whole brain homogenates from this patient though a marked effect was found with the use of whole brain homogenates from controls. The estimated "Triton®-activated" GDH activity was 10% of that found in controls. Determination of the enzyme activity in high-speed fractions prepared from whole brain homogenates revealed that the decrease in activity was limited to the particulate (membrane-associated) component (25% of control), with the soluble component showing normal activities. Thus, these results revealed that the same type of defect that was originally described in extraneural tissues of the patients with OPCA was also present in the brain of a patient dying from this disorder. These results are, therefore, consistent with a generalized defect in the membrane-associated form of GDH that is most likely genetically determined.

V. PURIFICATION AND CHARACTERIZATION OF BRAIN GDH

A. General

The study of human leukocytes and cultured skin fibroblasts provided evidence for the cellular presence of two forms of GDH differing in solubility and thermal stability characteristics in these tissues. To further explore whether similar forms of the enzyme are also present in mammalian brain, GDH activity was determined in high-speed fractions (100,000 xg for 60 min) obtained from whole rat brain homogenates after removal of the low-speed (480 xg for 10 min) pellet. Most of the high-speed enzyme activity was found to be particulate (membrane-associated). This particulate-bound activity resisted extraction by nonionic detergents, high concentrations of salt, and sonication; however, it was extractable with the cationic detergent, cetyl tetra ethyl ammonium bromide (CTAB).[4]

Readily solubilized and particulate GDH activities were purified by a combination of hydrophobic interaction, ion exchange, and hydroxy-apatite chromatography or by a combination of hydrophobic-interaction and GTP-Sepharose® affinity column. Both purified proteins displayed a similar molecular mass on sodium dodecyl sulfate-polyacrylamide gel electrophoresis (SDS-PAGE). However, they were found to differ significantly in kinetic constants (K_m for NADH and NADPH), allosteric regulation, and heat stability characteristics. The greatest differences were noted in their susceptibility to inhibition by GTP and haloperidol. Both agents inhibited the membrane-bound enzyme more efficiently than the soluble GDH, with apparent K_is of 18.5 and 9.0 μM GTP (in the presence of 1 m*M* ADP) and 23.5 and 6.5 μM haloperidol for the soluble and particulate enzymes, respectively. The Hill plot coefficient, binding constant, and cooperativity index (the soluble inhibited a higher degree of cooperativity) were significantly different, indicating that the two proteins differ in their allosteric sites.[4]

B. Subcellular Localization of Soluble and Particulate GDH Activities in the Rat Brain

GDH in the mammalian brain has been reported to be associated with mitochondria.[33] Within these organelles, the enzyme is thought to be localized in the matrix, being largely solubilized with tissue homogenization[34] and with Triton® X-100.[33] In contrast to this belief, our studies described above indicated that most of the GDH activity present in the rat brain is particulate-bound, not readily extractable by a variety of techniques expected to solubilize proteins located in the mitochondrial matrix. In an effort to determine the subcellular localization of the enzyme activities in the brain, we determined GDH activity in purified mitochondrial and synaptosomal preparations isolated from the rat brain.[41] In addition, succinate dehydrogenase (SDH) and NAD-isocitrate dehydrogenase (NAD-ICDH) were measured as mitochondrial markers and LDH as the cytosolic (synaptosomal marker). GDH activity was found to be twofold enriched over the two mitochondrial markers in the purified synaptosomes. Thus, the ratio of the mitochondrial/synaptosomal activities was 12.4:1 for SDH and 12.9:1 for NAD-ICDH, but 6.1:1 for GDH. LDH activity was present in the synaptosomal preparations, but it was absent in the mitochondrial preparations.

Most of the mitochondrial GDH activity (about 80%) was extractable by tissue disruption-homogenization and buffer washing (50 m*M* Tris HCl, pH 7.40). On the other hand, little (3 to 10%) of the synaptosomal activity was extractable by these techniques. CTAB could extract most of the synaptosomal GDH activity, but nonionic detergents and high-molarity salt solutions had little effect. When synaptosomal preparations were subfractionated into synaptic mitochondria and synaptic membranes, GDH activity was found to be readily solubilized from the mitochondria, but remained bound to the membranes. These results are consistent with the possibility that in the brain readily solubilized GDH is of mitochondrial matrix origin, whereas the particulate enzyme is associated with the synaptosomal membranes. It is, therefore, tempting to ascribe the former to the "metabolic pool" of glutamate and the latter to the "synaptic pool" of the amino acid.[42] The fact that the form of the enzyme that is associated with synaptic membranes is selectively decreased in patients with recessive OPCA raises important implications strongly supporting neuroexcitotoxic mechanisms in the pathogenesis of human degenerative disorders.

ACKNOWLEDGMENT

This work was supported by the National Institutes of Health (NIH) Grants NS-16871 and NS-1163, the Clinical Center for Research in Parkinson's and Allied Diseases, and the NIH Division of Research Resources Grant RR-71.

REFERENCES

1. **Kuhlman, R. E. and Lowry, O. H.,** Quantitative histochemical changes during development of rat cerebral cortex, *J. Neurochem.*, 1, 173, 1956.
2. **Bayer, S. M. and McMurray, W. C.,** The metabolism of amino acids in developing rat brain, *J. Neurochem.*, 14, 695, 1967.
3. **Leong, S. F. and Clark, J. B.,** Regional development of glutamate dehydrogenase from rat brain, *J. Neurochem.*, 43, 106, 1984.
4. **Colon, A., Plaitakis, A., Perakis, A., Berl, S., and Clarke, D. D.,** Purification and characterization of a soluble and a particulate glutamate dehydrogenase from rat brain, *J. Neurochem.*, 46, 1811, 1986.
5. **Fonnum, F.,** Glutamate: a neurotransmitter in mammalian brain, *J. Neurochem.*, 42, 1, 1984.

6. **Olney, J. W., Ho, O. L., and Rhee, V.,** Cytotoxic effects of acidic acid sulfur containing amino acids of the infant mouse central nervous systems, *Exp. Brain Res.*, 14, 61, 1971.
7. **Rothman, S.,** Synaptic release of excitatory amino acid neurotransmitter mediates anoxic neuronal death, *J. Neurosci.*, 4, 1884, 1984.
8. **Wieloch, T.,** Hypoglycemia-induced damage prevented by N-methyl-D-aspartate antagonist, *Science*, 230, 681, 1985.
9. **McGeer, E. G. and McGeer, P. L.,** Duplication of biochemical changes of Huntington's chorea by intrastriatal injection of glutamic and kainic acids, *Nature (London)*, 263, 517, 1976.
10. **McGeer, P. L., McGeer, E. G., and Singh, K.,** Kainate-induced degeneration of neostriatal neurons: dependency upon corticostriatal tract, *Brain Res.*, 139, 38, 1978.
11. **Coyle, J. T.,** Neurotoxic amino acids in human degenerative disorders, *Trends Neurosci.*, 5, 287, 1982.
12. **Bird, E. D. and Iverson, L. L.,** Huntington's chorea: post-mortem measurements of glutamic acid decarboxylase, choline acetyltransferase and dopamine in basal ganglia, *Brain*, 97, 457, 1974.
13. **Perry, T. L., Kish, S. J., Hansen, S., and Currier, R. D.,** Neurotransmitter amino acids in dominantly inherited cerebellar disorders, *Neurology*, 31, 237, 1981.
14. **Monfort, J. C., Javoy-Agid, F., Hauw, J. J., Dubois, B., and Agid, Y.,** Brain glutamate decarboxylase in Parkinson's disease with particular reference to premortem severity index, *Brain*, 108, 301, 1985.
15. **Desclin, J. C. and Escubi, J.,** Effects of 3-acetylpyridine on the central nervous system of the rat as demonstrated by silver methods, *Brain Res.*, 77, 349, 1974.
16. **Plaitakis, A.,** The olivopontocerebellar atrophies, *Semin. Neurol.*, 2, 334, 1982.
17. **Plaitakis, A., Nicklas, W. J., and Desnick, R. J.,** Glutamate dehydrogenase deficiency in three patients with spinocerebellar syndrome, *Ann. Neurol.*, 7, 297, 1980.
18. **Plaitakis, A.,** Abnormal metabolism of neuroexcitatory amino acids in the OPCAs, in *The Olivopontocerebellar Atrophies*, (Advances in Neurology Series, Vol. 41), Duvoisin, R. C. and Plaitakis, A., Eds., Raven Press, New York, 1984, 225.
19. **Langston, J. W., Ballard, P. A., Tetrud, J. W., and Irwin, I.,** Chronic parkinsonism in humans due to a product of meperidine-analog synthesis, *Science*, 219, 979, 1983.
20. **Herken, H.,** Functional disorders of the brain induced by synthesis of nucleotides containing 3-acetylpyridine, *Z. Klin. Chem.*, 6, 357, 1968.
21. **Plaitakis, A., Berl, S., Nicklas, W. J., and Yahr, M. D.,** Glutamate dehydrogenase deficiency in spinocerebellar degenerations: correlation with adult-onset recessive ataxia, *Trans. Am. Neurol. Assoc.*, 105, 476, 1980.
22. **Plaitakis, A. and Berl, S.,** Oral glutamate loading in disorders with spinocerebellar and extrapyramidal involvement: effect on plasma glutamate, aspartate and taurine, *J. Neural Transm. Suppl.*, 13, 67, 1983.
23. **Plaitakis, A., Berl, S., and Yahr, M. D.,** Neurological disorders associated with deficiency of glutamate dehydrogenase, *Ann. Neurol.*, 15, 144, 1984.
24. **Plaitakis, A.,** Biochemistry of recessive olivopontocerebellar atrophy, *Ital. J. Neural Sci. Suppl.*, 4, 65, 1984.
25. **Yamaguchi, T., Hayashi, K., Murakami, H., Ota, K., and Maruyama, S.,** Glutamate dehydrogenase deficiency in spinocerebellar degeneration, *Neurochem. Res.*, 7, 627, 1982.
26. **Duvoisin, R. C., Chokroverty, S., Leopore, F., and Niclas, W. J.,** Glutamate dehydrogenase deficiency in patients with olivopontocerebellar atrophy, *Neurology*, 33, 1311, 1983.
27. **Duvoisin, R. C. and Nicklas, W. J.,** Inheritance and expressivity of GDH deficiency, *Neurology*, 34 (Suppl. 1), 127, 1984.
28. **Duvoisin, R. C. and Plaitakis, A.,** The olivopontocerebellar atrophies, *Adv. Neurol.*, 41, 1, 1984.
29. **Schut, J. W. and Haymaker, W.,** Hereditary ataxia: pathological study of 5 cases of common ancestry, *J. Neuropathol. Clin. Neurol.*, 1, 183, 1951.
30. **Boller, F. and Segarra, J. M.,** Spinopontine degeneration, *Eur. Neurol.*, 2, 356, 1969.
31. **Bannister, R. and Oppenheimer, D. R.,** Degenerative diseases of the nervous system associated with autonomic failure, *Brain*, 95, 457, 1972.

31a. **Beaufay, H., Bendall, D. S., Baredhrem, P., and Duve, C. D.,** Intracellular distribution of some dehydrogenase; alkaline deoxyribonuclease and iron in rat liver tissue, *Biochem. J.*, 73, 623, 1959.

32. **Hodgeboom, G. H. and Schneider, W. C.,** Intracellular distribution of enzymes. XI. Glutamic dehydrogenase, *J. Biol. Chem.*, 204, 233, 1953.
33. **Salganicoff, L. and DeRoberties, E.,** Subcellular distribution of the enzymes of the glutamic acid, glutamine, γ-aminobutyric acid cycles in rat brain, *J. Neurochem.*, 12, 387, 1965.
34. **Fahien, L. A., Stremecki, M., and Smith, S.,** Studies of glucogenetic mitochondrial enzymes. I. A new method of preparing bovine liver glutamate dehydrogenase and effects of purification methods on properties of the enzyme, *Arch. Biochem. Biophys.*, 130, 449, 1969.
35. **Plaitakis, A., Colon, A. D., Perakis, A., Berl, S., and Clarke, D. D.,** Evidence for the presence of two forms of GDH in mammalian tissues. One is reduced in a neurological disorder, *Soc. Neurosci. Abstr.*, 9, 1171, 1983.

36. **Plaitakis, A., Berl, S., and Yahr, M. D.,** Abnormal glutamate metabolism in adult-onset degenerative neurological disorders, *Science*, 216, 193, 1982.
37. **Plaitakis, A. and Smith, J.,** Biochemical and morphological changes of brain in a patient dying of GDH-deficient olivopontocerebellar atrophy, *Ann. Neurol.*, 20, 152, 1986.
38. **Azzi, A., Campbell, J. B., and Robinson, B. H.,** Penetration of the mitochondrial membrane by glutamate and aspartate, *Biochem. Biophys. Res. Commun.*, 29, 148, 1967.
39. **Meijer, A. J. and VanDam, K.,** The metabolic significance of anion transport in mitochondria, *Biochem. Biophys. Acta*, 346, 213, 1974.
40. **Huang, Y. P. and Plaitakis, A.,** Morphological changes of olivopontocerebellar atrophy in computed tomography and comments on its pathogenesis, in *The Olivopontocerebellar Atrophies*, (Advances in Neurology Series, Vol. 41), Duvoisin, R. C. and Plaitakis, A., Eds., Raven Press, New York, 39, 1984.
41. **Plaitakis, A., Berl, S., and Clarke, D. D.,** Glutamate dehydrogenase in rat brain mitochondria and synaptosomes, *Trans. ASN*, Vol. 17, 218, 1986.
42. **Berl, S. and Clarke, D. D.,** The metabolic compartmentation concept, in *Glutamine, Glutamate and GABA in the Central Nervous System*, (Neurology and Neurobiology Series, Vol. 7), Hertz, L., Kvamme, E., McGeer, E. G., and Schousboe, A., Eds., Alan R. Liss, New York, 1983, 205.
43. **Young, A. B., Penney, J. B., Dauth, G. W., Bromberg, M. B., and Gilman, S.,** Glutamate or aspartate as a possible neurotransmitter of the cerebral cortico-fungal fibers in the monkey, *Neurology (Cleveland)*, 33, 1513, 1983.
44. **Young, A. B., Oster-Granite, M. L., Herdon, R. M., and Snyder, S. H.,** Glutamic acid: selective depletion in hamster cerebellum, *Brain Res.*, 73, 1, 1974.
45. **Brodal, A.,** *Neurological Anatomy in Relation to Clinical Medicine*, 3rd ed., Oxford University Press, Oxford, 1981.

Chapter 23

EXCITATORY AMINO ACIDS IN EPILEPSY

Henry F. Bradford and Dale W. Peterson

TABLE OF CONTENTS

I. BACKGROUND

A. General

Until recently a major effort in epilepsy research was directed toward investigating possible failures in GABAergic systems as a fundamental element in the underlying pathophysiology of epilepsy.[5,47] While an interest in this topic continues to thrive, there has been a change of direction by many research groups who have turned their attention toward excitatory amino acid systems. However, the possible involvement of endogenous excitatory amino acids as agents mediating seizures has a long history and dates back to the 1950s when Hayashi[1] (working in Tokyo, Japan) observed that inclusion of glutamate in fluids superfusing cerebral ventricles initiated seizures. During the intervening 25 years, researchers have clearly established the widespread occurrence of specific receptors for excitatory amino acids,[2,61] most of which probably interact with glutamate. These receptors are found in synaptic fields and are differentially distributed throughout the CNS in an anatomically distinct manner. This same period has seen the putative mapping of CNS neural pathways which use glutamate as their neurotransmitter and the realization that glutamatergic systems are ubiquitous and are likely to operate a very large proportion of the total synapses of the CNS.[3,4,48] These features qualify glutamate as a factor of special interest in the search for endogenous biochemicals which may be directly involved in basic mechanisms of epilepsy.[5] This involvement could be of two categories: (1) in the spread of seizure activity from its point of origin, via excitatory glutamatergic (or aspartatergic) neural pathways, and (2) in the generation of the primary abnormal hyperactivity at local sites.

B. Classification of Excitatory Amino Acid Receptors

The development of new analogues of excitatory amino acids has led to the detection of three subclasses of excitatory receptor in the brain and spinal cord.[7,16,123] These receptor subtypes, categorized by the electrophysiological potency of certain agonists, have become known as *N*-methyl-D-aspartate (NMDA), quisqualate, and kainate receptors, abbreviated A1, A2, and A3, respectively (Table 1). None of these compounds are present in neural tissue, but each is capable of inducing seizures as well as neurotoxic damage when injected into the brain,[62] as is glutamate, the endogenous nonselective agonist, which interacts with all three receptor subtypes.[59]

II. ROLE OF GLUTAMATE IN TRANSMISSION OF HYPERACTIVITY

The case for a likely role of glutamate, or of other excitatory amino acids, in the spread of epileptic seizures from the site of initial hyperactivity rests on the circumstantial fact that extensive glutamatergic pathways appear to course between various brain structures and between the cerebral cortex and many other subcortical brain regions.[6,7] This would provide the ready opportunity for seizure activity to spread from its site of origin to other brain centers and thereby initiate the various epileptic behavioral sequelae. Second, the potent anticonvulsant actions of specific receptor antagonists for excitatory amino acids (e.g., ω-phosphono-α-aminocarboxylic acids; see detailed discussion below) may in part be due to blockade of these widespread glutamatergic pathways, thus limiting the spread of hyperactivity.

III. ROLE OF GLUTAMATE IN PRIMARY EVENTS

A. Convulsions Induced by Glutamate

The evidence for glutamate as an excitatory factor in the primary events initiating epileptogenic hyperactivity has many threads. First, there is the convulsive action of glutamate when applied intraventricularly or by focal injection to various brain regions[8,59] or by systemic

Table 1
CLASSIFICATION OF EXCITATORY AMINO ACID RECEPTORS

	Subtype A1	Subtype A2	Subtype A3
Agonists	NMDA	Quisqualate	Kainate
	Ibotenate	AMPA	Domoate
	cis-ADCP		
	trans-PDA		
	Aspartate	Glutamate	Glutamate
	Glutamate	(Aspartate)	(Aspartate)
Antagonists	APH	GAMS	GAMS
	APV		
	cis-PDA	*cis*-PDA	
	Asp-AMP	Glutamyl taurine	Glutamyl taurine
	D-αAA	Glutamyl diethylester	
	γ-DGG	γ-DGG	
	Mg^{2+}		
	CPP		

Note: Scheme based on current data from electrophysiology and ligand-binding studies. *cis*-ADCP = aminodicarboxycyclopentane; *trans*-PDA or *cis*-PDA = 2.3- (or 2.4-) piperidinedicarboxylic acid; AMPA = D-amino-3-hydroxy-5-methyl-4-isoxazolepropionic acid; Asp-AMP = β-D-aspartylaminomethylphosphonate; D-αAA = D-α-aminoadipic acid; γDGG = γ-D-glutamylglycine; GAMS = γ-D-glutamylaminomethylsulfonate; NMDA = *N*-methyl-D-aspartate; APH = aminophosphonoheptanoate; APV = aminophosphonovalerate; and CPP = 3-[(±)-2-carboxypiperazin-4-yl] propyl-1-phosphonic acid.

injection.[9-11] The effectiveness of the blood-brain barrier in most brain regions prevents glutamate in the blood from gaining access to the brain unless blood levels are high.[11] In young animals (e.g., 10-day-old rats), where the blood-brain barrier is immature, much lower levels of blood glutamate precipitate convulsion.[9]

B. Excitotoxic Action

Glutamate given systemically can cause tissue lesions in the circumventricular organs of the brain where there is effectively no blood-brain barrier due to the presence of fenestrated capillaries.[12,14] These so-called excitotoxic actions of glutamate which can be also induced by other excitatory amino acids appear to be receptor-mediated since in many cases they can be prevented by receptor antagonists.[13,16] The excitotoxic tissue damage caused by glutamate, and more powerfully by its congeners, is characteristic in that it is particularly directed against dendrites and spine processes causing dendritic swelling together with some swelling of local glial processes. In contrast, nerve terminals synapsing on the dendrites are largely unaffected. These dendrites are assumed to be postsynaptic to glutamatergic neurons. Subsequently, many neurons local to the injection site degenerate (e.g., intrinsic GABAergic or cholinergic neurons in striatum), while the nerve terminals and axons of extrinsic neurons projecting to the injection site, as well as local glial cells, seem to survive. Hence, the term "axon-sparing" is sometimes applied to the degenerative actions of excitatory amino acids.

The processes leading to excitotoxic neuronal death have not been precisely determined, but are probably connected with excessive chronic, receptor-mediated, depolarization of the postsynaptic structures. This would lead to low levels of ATP and phosphocreatine due to accelerated Na^+ pump activity and to restricted activity of various vital energy-requiring processes, as has been shown to occur in brain slices treated with excitatory amino acids.[50] It is likely that excessive Na^+ entry is the major initial ionic event,[50] and subsequent exaggerated uptake and accumulation of Ca^{2+} or of chloride ion could be an important associated event leading to neuronal or glial cell death[51,130,131,133] (see below). Other excitatory amino acids (e.g., kainate, ibotenate, NMDA, homocysteate, and quinolinic acid) are much more effective than glutamate itself in producing excitotoxic lesions,[14] but both their mode of action and the nature of the lesions they produce seem to be essentially similar.

IV. NEUROTOXIC LESIONS AS EPILEPTIC FOCI

A. General

Kainate injected into the striatum usually causes hippocampal neurodegeneration as well. The necrotic lesions produced can become chronically epileptogenic and may produce spontaneous recurrent seizures in rats.[49,97] The pattern of hippocampal degeneration due to kainate was similar to that commonly seen in temporal lobe epilepsy in man, as it involves neuronal losses in the CA1 and CA3 hippocampal fields, with sparing of both the CA2 field and the dentate gyrus.[49,88] Clearly, if endogenous glutamate (or some unidentified endogenous excitatory amino acid) were to cause lesions due to such an excitotoxic action, these too could have the capacity to become epileptogenic. The high tissue concentrations of free glutamate in brain tissue (10 m*M*) would particularly qualify glutamate for an excitotoxic role of this kind. Nonspecific changes in membrane permeability to soluble cytosolic constituents would readily result in a local acute or chronic accumulation of glutamate extracellularly due to the high outwardly directed, diffusion gradient if the efflux rate was greater than the high-affinity (low V_{max}) uptake of glutamate could cope.[5] Massive spillover of glutamate to the extracellular space could be expected to occur during brain damage due to head trauma or even following brain surgery, both of which can result in epilepsy.[52,53] Though many experimentally induced lesions do not become epileptogenic,[94] brain scar tissue formation or gliotic changes associated with tumors, head injury, infection and meningeal adhesion, hematoma, or hypoxic cell death necessarily result in changed tissue organization (e.g., changed glia to neuron ratio, presence of necrotic regions, and local failure of the blood-brain barrier) and often become epileptogenic.[5,53] Such structural or cellular changes could result in an alteration in the balance of enzyme activities related to glutamate function or local entry of glutamate from the blood. Changed tissue organization which leads to a less effective contribution by high-affinity glutamate uptake to the clearing of extracellular glutamate could lead to a rise in its extracellular level to a point where it was able to cause direct excitation during excessive glutamatergic pathway activity. Similarly, such excessive activity might itself lead to excitotoxic lesion or to another more subtle category of "kindling" change in the brain. For instance, "mirror foci" which commonly develop at points contralateral and homotopic to active experimental foci may well be generated by such changes.[5,60]

The relatively low potency of glutamate in causing neurodegeneration compared with kainate and other analogues has been attributed to its removal by high-affinity inward transport. When the relatively specific glutamate transport inhibitor, D,L-threo-3-hydroxy-aspartate, was injected intrastriatally, it caused extensive neurodegeneration and potentiated the action of coinjected glutamate.[54] The action of the uptake inhibitor alone in the striatum was substantially reduced by prior decortication which causes removal of the glutamatergic corticostriate pathway. This suggests that the reduced availability of glutamatergic terminals (and therefore of glutamate for release)[126] was the reason for the diminished action of the uptake inhibitor.[54]

There is evidence that the neurotoxic actions of kainate in hippocampus and striatum are dependent upon, or augmented by, the presence of afferent glutamatergic nerve terminals.[54-56] The nature of this interaction between kainate and glutamate remains obscure, but may be due to enhanced glutamate release from its terminals due to an action of kainate[54] (e.g., on presynaptic receptors;[57,96] see below).

Whether endogenous glutamate itself* can cause excitotoxic lesions in hippocampus or elsewhere which could become epileptogenic (e.g., kainate)[49,97] remains in the area of speculation and has not yet been investigated. Lesions due to injections of an uptake blocker (e.g., D,L-3-hydroxyaspartate)[54] into hippocampus, or amygdala, with or without exogenous glutamate needs to be examined (over long periods) for epileptogenic properties.

The relationship between kainic acid-induced lesions in hippocampus and recurrent seizures has been the subject of several studies. It seems that distant lesions of the septum, claustrum, and the contralateral cortex and hippocampus result from the seizures themselves. This was shown by blocking the kainate-induced seizure with diazepam, which resulted in the lesions being limited to the hippocampal injection site.[89] A more recent study concluded that the acute histopathology induced in hippocampal pyramidal cells by kainic acid is also a result of the induced seizures.[90] Experimental lesions of the amygdala prior to the injection of kainic acid decreased the recurrent convulsions and the seizure activity in the hippocampus, suggesting that amygdala-hippocampal interaction is important in maintaining the seizure activity. Despite the reduced severity of seizures in these animals, loss of the pyramidal cells was still as evident, suggesting that a direct cytotoxic action of kainate was responsible.[91]

The autoradiography of excitatory amino acid receptor subtypes has revealed their localized anatomical distribution, with kainic acid A3 (Table 1) receptors specifically distributed to the CA3 region.[2] Electrophysiological evidence[92] and glucose utilization studies[93] both indicate a high degree of localization of kainic acid excitation to the CA3 region. Since the CA3 region projects to CA1 pyramidal cells, the basis for a direct action leading to kainic acid lesions in the hippocampus exists. There remains the question of whether local excitation is sufficient to induce lesions or whether some recurrent pathway is involved which feeds back excitation into the hippocampus to the degree necessary to cause cell death.

Electrophysiological measurements in hippocampal slices have shown that kainic acid depolarization may be due both to direct actions on A3 receptors (slow response) and to an indirect action due to released glutamate or aspartate[95] (fast response). Measurements of kainic acid-induced release of amino acids from brain slices[19,96] indicate that the indirect action is due to a specific A3 (presynaptic) receptor on excitatory terminals in a number of brain regions.[57,96] This release was shown to be Ca^{2+} dependent and was not blocked by tetrodotoxin (TTX).[96] Other agonists, with potent excitatory effects (e.g., NMDA and ibotenic acid), did not release glutamate or aspartate. These observations all point to a specific presynaptic action of kainate which could account for the potent and selective neurotoxic action of this compound. Kainic acid may therefore be a useful tool for the study of lesions associated with epilepsy as it may be activating an endogenous mechanism which can lead to a specific category of neuronal cell impairment or death mediated by excessive release of glutamate.

The important question that remains about kainate-induced recurrent seizures is their relation to the neurotoxic effects. Correlation of the long-term effects with epileptiform activity in hippocampal slices showed that the loss of after hyperpolarization and inhibitory postsynaptic potentials in CA1 was due to the loss of the CA3 and CA4 cells at 2 to 4 weeks

* Recently shown to be so (Mori, N. and Wada, J. A., *Brain Res.*, 425, 145, 1987.)

after injection.[97] While the cell loss was permanent, these physiological correlates returned to normal after 2 to 4 months when the seizures no longer occurred.

B. Quinolinic Acid

An endogenous excitotoxin with a profile of activity very much like kainic acid is quinolinic acid. It is more potent as a neurotoxin than it is as an excitant and seems to require intact excitatory inputs to produce its specific lesions. The immediate excitatory effects of quinolinic acid are blocked by A1 receptor antagonists,[98] but the convulsions induced by injection of this agonist are not blocked by APH (aminophosphonoheptanoate), which does completely inhibit NMDA convulsions.[122] However, antagonism of quinolinic acid-induced excitation was measured only a few minutes after application of the agonists, while convulsions occur with a latency of 20 min after i.p. injection. A sufficient time of 20 min would allow for metabolism of quinolinic acid to one of the many derivatives of quinolinic acid which have convulsant activity,[99] perhaps mediated by A2/A3 receptors[100] (Table 1).

If quinolinic acid or one of its metabolites is involved in generating epileptic seizure activity, some mechanism by which the very low brain levels of these compounds are raised would need to be identified. Evidence to date suggests that quinolinic acid does not display the properties of discrete release and rapid inactivation required for a neurotransmitter.[101] How quinolinic acid could be released and excite sufficient neurons to precipitate a seizure therefore remains to be explained.

C. Ischemic Brain Damage

Glutamate and aspartate have also been identified as the agents responsible for the neuropathological damage caused during ischemia and hypoxia, since release of these compounds occurs when such conditions are created experimentally in vivo.[15] Also, ischemic tissue damage can be mimicked by glutamate and aspartate added to neuronal cultures.[16] The effects of these excitatory amino acids appear to be mediated via receptors in both situations.[16,17] Presumably, the severe depolarizing actions of glutamate and aspartate induce neuronal death in ischemia and in anoxia by an excitotoxic action. Excessive entry of extracellular Ca^{2+} under these conditions may be a factor in causing cell death,[51] possibly by causing exaggerated activity of Ca^{2+}-dependent proteases or phosphorylations. Glutamate, along with kainate and NMDA, has been shown to cause enhanced entry of Ca^{2+} to hippocampal and also brain slices in vitro,[18,19] and this, too, can be prevented by excitatory amino acid receptor antagonists[19] (Table 2). Also, Ca^{2+} entry induced by the depolarizing agents veratrine and by high K^+ can be partly blocked by these receptor antagonists, suggesting that a large part of the effect is mediated by released glutamate and aspartate[19] (Table 2). The very high levels of these free amino acids in the cytoplasm of both neurons and glial cells (3 to 10 m*M*) provide a plentiful source for release during hypoxia or ischemia when lowered production of energy phosphates would limit glutamate transport processes and allow its accumulation in the extracellular space. This would be in addition to glutamate being released by synaptic activity, though one recent study suggests that ischemia specifically enhances the extent of Ca^{2+}-dependent synaptically released glutamate.[44]

Others have reported that while Ca^{2+} entry may be involved in longer-term excitotoxic mechanisms,[130,132,133] chloride ions passively influxing during depolarization are more directly involved in acute glutamate-induced cell death, probably due to osmotic changes with a large increase in cell water content. Such processes appear to be Ca^{2+} independent and Cl^- dependent.[130-132,134] Indeed it may be glial cells which first deteriorate as a result of neuronal overexcitation and consequent exaggerated ionic influx (Cl^-, K^+) to these cells with accompanying osmotic swelling and lysis.[131]

Table 2
BLOCKADE OF DEPOLARIZATION OR AMINO ACID-INDUCED $^{45}Ca^{2+}$ ENTRY TO HIPPOCAMPAL SLICES BY EXCITATORY AMINO ACID ANTAGONISTS

	$^{45}Ca^{2+}$ uptake (nmol/min)		
Agent	**Control**	**Stimulated**	**Stimulated + AP7 (500 μ*M*)**
Veratrine (75 μ*M*)	1764 ± 114	3225 ± 182	2164 ± 79
K^+ (56 m*M*)	1764 ± 114	3159 ± 118	2049 ± 141
NMDA (50 μ*M*)	2062 ± 127	2722 ± 37	1812 ± 272
Kainate (50 μ*M*)	2671 ± 94	3648 ± 198	3655 ± 149
Glutamate (100 μ*M*)	2518 ± 121	3122 ± 145	2781 ± 91

Note: Hippocampal slices were incubated at 37°C in Krebs-Ringer's-Tris containing glucose (10 m*M*) and $CaCl_2$ (1.2 m*M*). After a 30-min preincubation, $^{45}CaCl$ was added and 1 min later depolarizing agents were added. Incubation ceased after a further 11 min. After rinsing in EGTA stop solution, ^{45}Ca was measured in the dissolved slice. Data are mean ±SEM (n = 6).

From Crowder, J., Croucher, M., Bradford, H. F., and Collins, J. F., *J. Neurochem.*, 48, 1917, 1987. With permission.

V. GLUTAMATE AND SPREADING DEPRESSION

The strong excitatory action of topically applied glutamate can paradoxically also lead to inhibition due to onset of spreading depression. This condition manifests itself as a general neuronal inhibition and loss of spontaneous electrical activity. It spreads slowly (3 mm/min) in all directions from the initial point.[20] It is propagated by the release of potassium and glutamate in the advancing front. This causes further depolarization and thereby triggers further release. The neurons recover their membrane potentials in about 2 min, but the depression of excitability lasts for about 10 or 15 min. The initial depression is due to the excessive depolarization which prevents recovery of the neuronal membrane potential sufficiently to allow generation of action potentials. The longer-lasting inhibition is likely to be due to GABA release.[20] A brief burst of action potentials is often recorded in the advancing front of spreading depression due to transitory activation of the neurons as the membrane potential passes through the critical firing level before gross and chronic depolarization causes depression of activity. Indeed there are similarities between spreading depression and the paroxysmal depolarization shifts (PDS) seen in many experimental epileptic foci. These too involve a massive and long-lasting depolarization accompanied by discharge of action potentials which is usually followed by inhibition of activity due to excessive depolarization. The phenomenon of spreading depression is much easier to evoke in the cerebral cortex of small mammals such as the rat and rabbit with smooth cortices and a high neuronal population density than in cats and primates which have highly folded cortices with many sulci and lower cell densities. Spreading depression has yet to be demonstrated in man. Thus, in primates and man it may not be possible for glutamate to set in train the events which lead to spreading depression.[21] The extent to which the actions of glutamate in spreading depression are receptor mediated has not been investigated, though the newly available specific antagonists now make such studies possible.

VI. EXTRACELLULAR ACCUMULATION OF GLUTAMATE IN EPILEPTOGENIC TISSUE

A. General

From the foregoing discussion, it is clear that glutamate and aspartate release from neural cells and/or their accumulation in the extracellular space can lead to a number of excitatory phenomena, many of which appear to be receptor mediated. These include excitation and seizure, excitotoxic neuronal death, ischemic and hypoxic neuronal death, and spreading depression and PDS[114] (see previous section).

The possibility that glutamate accumulating extracellularly could trigger epileptic hyperactivity has been under consideration for some 15 years, and a good deal of data from experimental animal models of epilepsy have accumulated in support of this hypothesis.[5,38]

B. Glutamate Release from Active Epileptic Foci

The earliest report of this phenomenon described glutamate release to cortical superfusion cups placed over active cobalt(Co-)-induced foci in the cerebral cortex of anesthetized cats.[22] The time course of glutamate release corresponded with the onset of epileptic hyperactivity measured by EEG. Examination of the amino acid content of the excised focus showed that the contents of glutamate, aspartate, GABA, and taurine were reduced by 30%. Other studies of changes in the levels of these amino acids in excised cobalt-induced epileptic foci have reported similar decreases in levels of these compounds and have revealed a correlation between the extent of these losses and the severity of the epilepsy.[23,24] The release of glutamate from active Co-epileptic foci induced in the sensorimotor area of rat cortex has been detected in awake, unrestrained animals by the use of a chronic cortical superfusion cup.[25] The release was specific to glutamate and correlated well with the onset and cessation of contralateral myoclonic limb jerks. Moreover, implantation of nickel, which causes similar scarring which is not epileptogenic, did not result in a similar pattern of glutamate release.[58] Other forms of experimental epilepsy have shown similar changes in amino acid content and a propensity to release glutamate. These include the decreases in glutamate, aspartate, GABA, and taurine observed in cerebral cortex of cats made epileptic by i.m. injections or topical application of penicillin.[26,27] Indeed, penicillin releases glutamate from cortex at concentrations that induce epileptiform activity.[118] Freeze-induced foci show a similar combination of decreased amino acid content and increased glutamate release.[5] In all cases, these changes paralleled the progressive increase in excitability of the cortex.[26] There were similar findings in the excised cortical foci of a cat with spontaneous grand mal epilepsy.[28]

In another study, the content and release of glutamate from cerebral cortex of the cat made epileptic by chronic undercutting were investigated.[29] Again, specific losses of glutamate, aspartate, and GABA occurred as the undercut cortex developed epileptiform electrical afterdischarges. When the epileptic cortex was electrically stimulated, it showed a threefold increase in glutamate release, whereas intact cortex displayed little change in this rate. Although other amino acids (GABA and aspartate) showed amplified release in response to stimulation, in *absolute* terms the increased glutamate release was tenfold greater.[29]

VII. LEVELS OF AMINO ACIDS IN EPILEPTOGENIC TISSUE

In the same epileptogenic cortical tissue samples where glutamate, aspartate, GABA, and taurine were decreased, a range of other amino acids were found to be increased. The most consistent and significant increases were in glycine, serine, and threonine, with glutamine and taurine being less consistent in their changes. The reason for these increases is unknown.

It has been suggested that the reduction in levels of glutamate, aspartate, and GABA found in epileptogenic cortex from experimental models may be due to changes in amino acid metabolism. These amino acids are important intermediates in mainstream carbohydrate and energy metabolism, in addition to their role in protein and peptide synthesis, and other biosynthetic pathways. The greater part of the content of these amino acids in neural tissue would be concerned with these metabolic activities rather than neurotransmission (except, perhaps, for GABA). For this reason it is important, when attempting to connect changed amino acid levels with seizure initiation or propagation, that aberrations in these processes are given equal consideration alongside any direct influence these amino acids may have on excitation or inhibition.

Restricted availability of glucose, or of oxygen, in hyperactive tissue could lead to the observed fall in pool size of glutamate, aspartate, and GABA, since Krebs' cycle intermediates would be depleted and there would be a decreased incorporation of carbon into these amino acids. However, slices of human temporal lobe and slices of the spiking cortical focus from Co-epileptic rats show no obvious change in respiratory rates,[80] and the more relevant in vivo findings show glucose uptake and incorporation to be *increased* during seizures rather than decreased.[30,31]

Although there are likely to be some changes in the rates of production and utilization of these amino acids which are linked to the enhanced metabolic response of the hyperactive tissue, this seems inadequate to explain the extent and pattern of the observed decreases in pool size. The several superfusion studies discussed above suggest that release of glutamate from various epileptic foci could account at least in part for the fall in pool size of this amino acid. In most cases the pattern of release showed a specificity toward glutamate. This would not be inconsistent with the observed multiple decrements in tissue amino acids, since a reduction in tissue glutamate content due to efflux would be reflected as parallel reductions in both GABA and aspartate, for both of which it is a precursor.

Analysis of amino acid content of excised human temporal lobe foci removed during surgical treatment for epilepsy have revealed alterations in content similar to those found in experimental animal models. Thus, glutamate, aspartate, GABA, and taurine were decreased, and serine, threonine, and glycine were increased in level.[32,33]

Quite why glutamate should accumulate extracellularly and diminish in level intracellularly is not clear. This could result from excessive activity of glutamatergic neurons (possibly due to loss of inhibitory control) or to a change in the permeability properties of neuronal plasma membranes leading to efflux of cytoplasmic glutamate under certain conditions. The third possibility is that glutamate accumulates extracellularly due to reduced levels of activity of the high-affinity transport system. This normally keeps these levels very low by inward transport of glutamate continuously leaking from various cell regions at a low rate and by release due to synaptic activity. Such a scheme has been previously suggested.[121]

Several other lines of circumstantial evidence implicate glutamate excitation in the initiation or spread of epilepsy.[5,35] Included here is the finding that several well-established anticonvulsants (e.g., phenytoin, phenobarbitone, diazepam, and trimethadione), at concentrations which would be anticonvulsant in vivo, can inhibit the stimulated release of excitatory amino acids in vitro,[36,37,43,59] as well as diminish inward calcium fluxes[43] (see below).

VIII. EXCITATORY AMINO ACID ANTAGONISTS AS ANTICONVULSANTS

The most direct demonstration of the likely involvement of glutamate (or other excitatory amino acids) in epileptic processes came when some of the newly developed specific receptor antagonists for these compounds were found to be potently anticonvulsant against experimental epilepsy. In the first published study employing this strategy,[38] D,L-2-amino-4-phosphonobutyric acid (D,L-APB) and D,L-2-amino-5-phosphonovalerate (D,L-APV) superfused

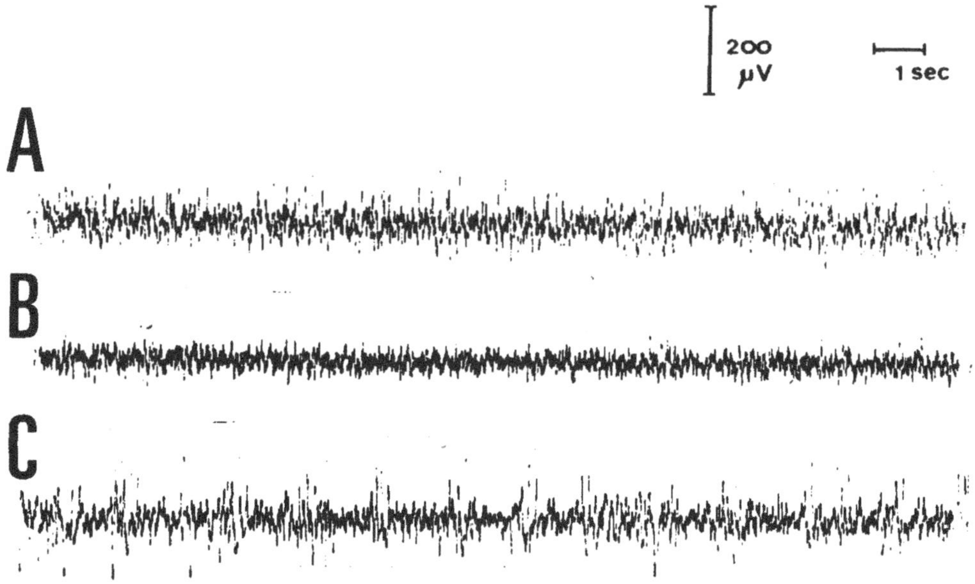

FIGURE 1. Electrocorticogram of sensorimotor cortex of an awake, unrestrained rat with Co-induced epilepsy. The cobalt focus was directly superfused with artificial CSF via a chronically implanted cortical cup. Recordings were made from electrodes implanted beside the cup and above the sensorimotor region. (A) Before drug addition; (B) 120 min after addition of D,L-APB or D,L-APV (0.55 m*M*); and (C) 10 min after subsequent superfusion with drug-free CSF.[38]

at low concentration (0.05 to 1.0 m*M*) directly onto active cortical co-epileptic foci in awake, unrestrained rats, totally prevented both contralateral myoclonic limb jerks and epileptiform EEG[38] (Figure 1). Subsequently, these and another homologue in the series, 2-amino-7-phosphonoheptanoate (D,L-APH) were tested against other animal models of epilepsy. Both D,L-APV and the new homologue were found to be active against chemically induced convulsions and genetically determined audiogenic seizures in mice.[39] Later, other categories of experimental seizures including photogenic epilepsy in baboons[46] and kindled amygdala or hippocampus in rats[41,42] were found to be controlled by APB, APV, or APH, respectively.

So far, the seven-carbon member of the phosphono-homologue series, APH, has proved to be the most widely effective anticonvulsant and is effective when given systemically,[39] as it appears to cross the blood-brain barrier due to its hydrophobic nature. It is known to be a specific antagonist of A1-type excitatory amino acid receptors which interact with NMDA (see details below). This suggests that NMDA receptors (A1; Tables 1 and 3) could strongly feature in the excitatory mechanisms generating a wide range of experimental seizures. However, the anticonvulsant actions of D,L-APB and GAMS (γ-D-glutamylaminomethylsulfonate) show that other receptor subtypes (A2 and A3; Tables 1 and 3) may also be playing a critical part in the generation of epilepsy.[38,77,124] It is notable, therefore, that the brain region in man and other mammals which contains the greatest concentration of NMDA receptors is the Sommer section (i.e., CA1 area) of the hippocampus.[139,140] This is the region where major cell loss occurs[5] in the most common form of human epilepsy (i.e., temporal lobe epilepsy).

The advent of these new and specific excitatory receptor antagonists has and effect relationships between glutamate release from experimental foci and the onset of epileptic hyperactivity to be probed further. The previously established coincidence of these two phenomena[5,22,25,41,45] did not rule out the possibility that glutamate release occurred to an

Table 3
EVIDENCE FOR SPECIFIC AMINO ACID RECEPTOR SUBTYPES IN VARIOUS EPILEPSY MODELS

Receptor subtype	A1	A2	A3
Epilepsy models	NMDA seizures (62)	Quisqualate seizures (62)	Kainate seizures (62)
	Audiogenic seizures (39)	Cobalt focus (38)	Audiogenic seizures (77)
	PTZ seizures (39)	Hippocampal kindling (42)	Picrotoxin seizures (78)
	Photosensitive baboon (40)	Homocysteine seizures (73)	Electrical kindling (11)
	Hyperbaric oxygen (63)	Quinolinic seizures (100)	
	Amygdala kindling (41)	Angular bundle kindling (110)	
	Spinal seizures (114)		

Note: Various animal models of epilepsy are grouped according to the excitatory amino acid subtype likely to be involved in the initiation of the seizures. This is based on the anticonvulsant action of specific receptor antagonists or differential changes in the receptor populations. Numbers in parentheses are references. NMDA = *N*-methyl-D-aspartate; and PTZ = pentylenetetrazole.

excessive extent only *after* onset of seizure activity and was not therefore responsible for the onset of this activity, though it could aid its prolongation.

The finding[41,42] that the phosphono derivatives could block the after discharges of kindled epilepsy recorded at the kindled site (Figure 2) indicates that the drugs were preventing the primary events *initiating* epilepsy, presumably by receptor blockade in this region. This does not preclude any additional action they might have in preventing the spread of activity from this focal area to other brain regions, with precipitation of the various behavioral components of epilepsy, including full grand mal seizures.

These findings could be interpreted as an action of the specific glutamate receptor antagonists at the focus in (1) preventing the excitatory effect of glutamate accumulating extracellularly (Figure 3), (2) preventing the generation of focal hyperactivity due to some other cause such as loss of local inhibitory neurons (e.g., GABAergic,[46] Figure 4), where such activity was dependent on transmission through glutamatergic (or aspartatergic) neurons, and (3) inhibiting synaptic release of glutamate and aspartate.[19]

IX. EXCITATORY AMINO ACID RECEPTORS MEDIATING SEIZURES

Audiogenic seizures and pentylenetetrazole-induced seizures in mice have been employed to compare the anticonvulsant potency of a number of these analogues of dicarboxylic amino acids.[39] The most potent in these particular models was APH. The potency of other phosphono analogues as well as less specific receptor antagonists paralleled their A1 receptor binding activity. Further studies used a wider range of seizures models such as photosensitive baboons,[40] other chemical convulsants,[122] and seizure induced by hyperbaric oxygen.[63] A larger range of antagonists upheld the correlation with A1 receptor activity including a series of piperidine dicarboxylic acids which show both anticonvulsant and proconvulsant actions[64] and two dipeptides which were potent A1 antagonists[65] (Table 3).

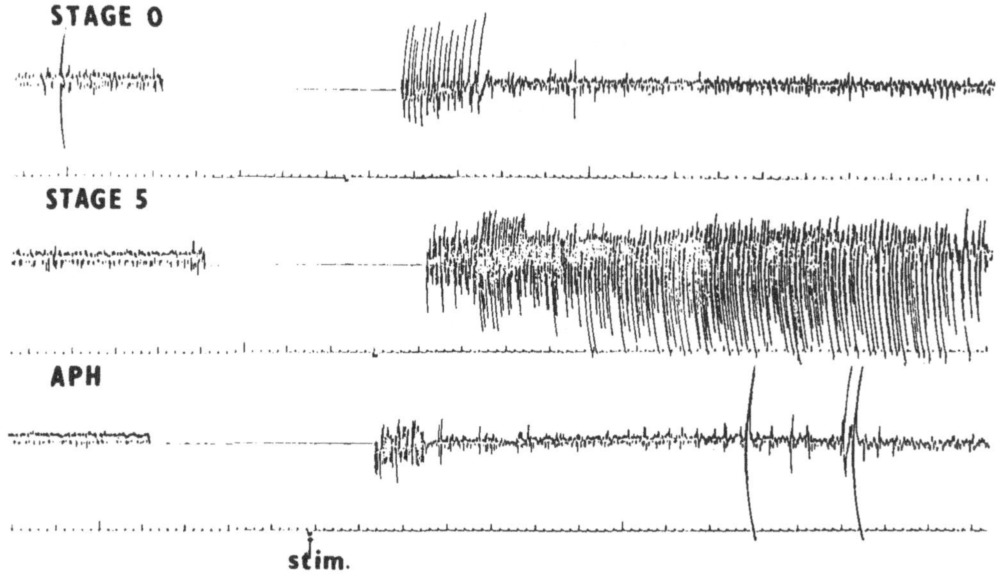

FIGURE 2. Afterdischarges recorded from amygdala-kindled rat showing effect of APH given intraventricularly. Stage 0 refers to kindled state after one electrical stimulus. Stage 5 refers to fully kindled state (i.e., after 12 daily stimuli). The bottom trace shows the absence of afterdischarge 45 min following intraventricular administration of APH (0.5 μmol in 10 μℓ of saline) to a rat kindled to stage 5. The electrode was positioned in the basal lateral nucleus of the amygdala.

A number of amino acid antagonists without selective or potent action at this receptor have been shown to be effective in other epilepsy models. As described above, D,L-APB and D,L-APV were shown to be anticonvulsant in the kindling model of epilepsy which, like the cobalt focus, was used to measure the antagonism of the local electrophysiological events at the focus as well as the overall convulsive components of the seizures. APB was far less active than APH or the closely related A1 antagonist APV on kindled amygdala seizures and the associated electrical afterdischarges.[41,45] In contrast, kindled seizures arising in the hippocampus (which are of lesser severity) were blocked completely by APB in some animals.[42,45] This action of APB in hippocampal kindling is likely to be related to the known suppression of synaptically mediated excitation in the hippocampus, which is caused by this antagonist.[66]

While the binding and displacement from A1 receptors and antagonism of NMDA excitation have presented a consistent explanation for the anticonvulsant actions of APV and APH, the relevance of APB receptor binding is less clear. Examination of the displacement of ^{3}H-APB by various analogues suggests that quisqualate (or A2) receptors may be specifically blocked by APB. The ionic requirements of ^{3}H-APB binding, which correspond to that portion of ^{3}H-glutamate binding which is dependent on Cl^- and Ca^{2+}, have suggested that APB identifies a separate, fourth class of excitatory amino acid receptor,[68] with a distinct anatomical distribution in the hippocampus[69] (Table 1). A range of APB analogues failed to produce a correlation between displacement of binding to this Cl^-/Ca^{2+}-dependent receptor and synaptic depressant activity.[70] These discrepancies between binding and function for this Cl^-/Ca^{2+}-dependent site may be due to the presence of a Cl^--dependent uptake system.[71] Thus, vesicular structures may be present which could transport and accumulate glutamate. However, the synaptic activity of APB is not readily explained by its blockade of glutamate uptake. Further complicating interpretation of the anticonvulsant actions of D,L-APB are the varied activities of the stereoisomers of this glutamate analogue. Thus, L(+)-APB is a synaptic depressant, probably due to blockade of the release of an excitatory

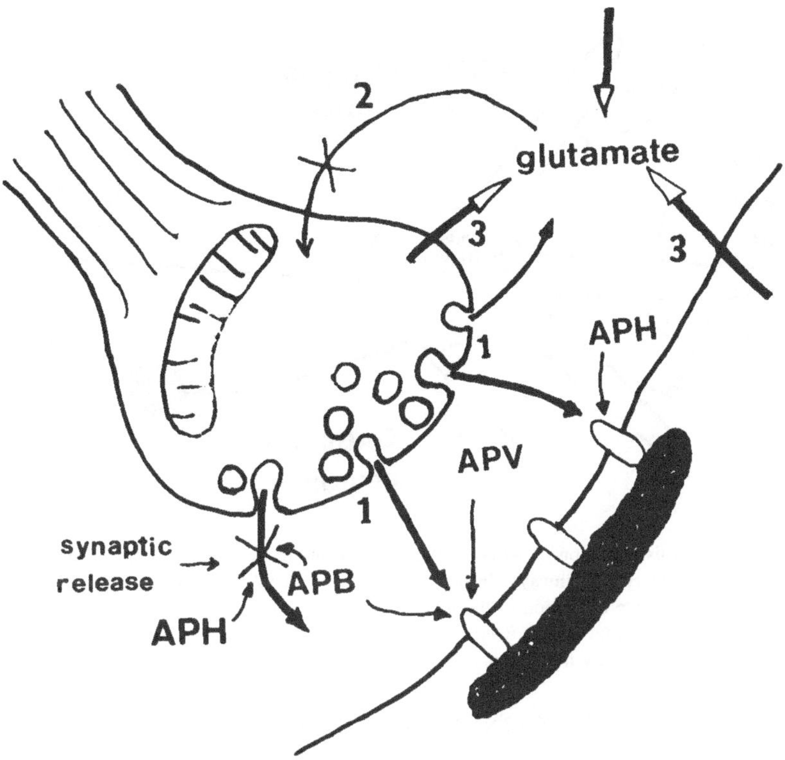

FIGURE 3. Proposed action of endogenous glutamate in generating epileptic hyperactivity. Glutamate (or aspartate) accumulating extracellularly as a result of (1) excessive synaptic activity, (2) reduced effectiveness of high-affinity uptake, or (3) leakage through membranes acts on postsynaptic excitatory amino acid receptors, the subtype (A1 to A3) depolarizing on the region involved. APH, APV, or APB prevent these actions at pre- or postsynaptic sites.

transmitter.[72] It is also a weak A1 agonist, while the D(−)-isomer is a weak A1/A2 antagonist.

Other attempts to demonstrate the involvement of A2 receptors in the mediation of epileptic seizures have used the A2 agonist homocysteic acid together with glutamate diethyl ester (GDEE), a nonselective A2 antagonist. Homocysteic acid-induced seizures were blocked by a dose of GDEE which had no effect on pentylenetetrazole seizures.[73] The recently synthesized kainate analogue, β kainic acid, could be expected to be a selective A3 (kainate) receptor antagonist. While β kainic acid does have anticonvulsant activity against audiogenic seizures,[74] this action may not be due to activity on A3 receptors.[75] Stone and Collins[76] were unable to demonstrate any antagonist action of β kainic acid at the electrophysiological level in cortical or hippocampal slices, but have shown a weak agonist action, which they suggest may mediate anticonvulsant activity by a presynaptic mechanism.

A potent and somewhat selective antagonism of A3 receptors has been more convincingly shown by GAMS. As an anticonvulsant, GAMS is more potent against kainate-induced seizures than NMDA or quisqualate-induced seizures.[124] GAMS is also a potent anticonvulsant in the DBA/2 audiogenic mouse,[77] an epilepsy model shown to be sensitive to A1 antagonists, indicating that in this genetic form of epilepsy and probably in most animal models, convulsions may be mediated by more than one excitatory receptor[72] (Table 3).

Multiple receptor function in the generation of seizure activity is also suggested by the potent antagonism of picrotoxin-induced convulsions by 3-hydroxy-2-quinoxalinecarboxylic acid (HQC) which is active at both A1 and A3 receptors.[78]

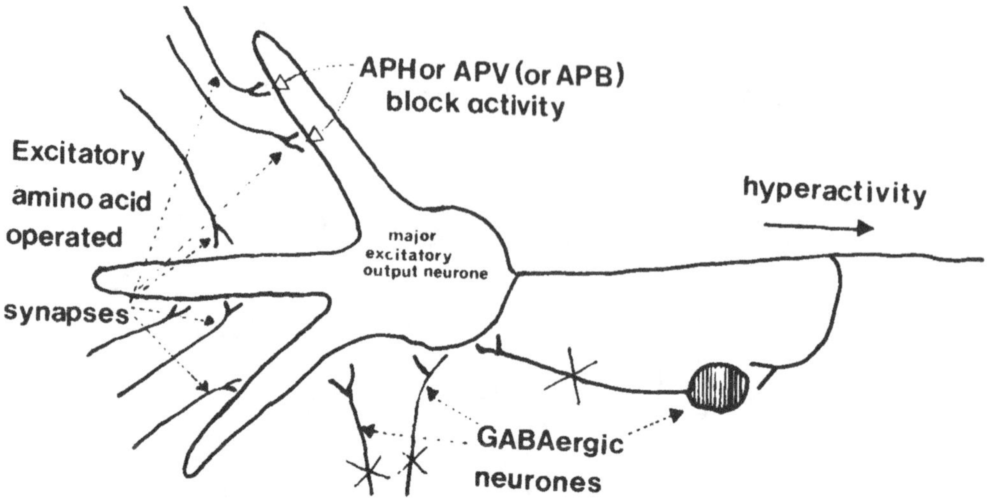

FIGURE 4. Anticonvulsant action of APH, APV, or APB due to prevention of spread of hyperactivity via glutamatergic (or aspartatergic) pathways. In the model presented, inhibitory GABAergic neurons are shown as failing, allowing excessive excitation of the main output neuron by glutamatergic inputs. This is prevented by antagonists. Where the output neuron itself is glutamatergic, transmission of the hyperactivity to other brain regions will also be prevented by the antagonists.

X. ESTABLISHED ANTICONVULSANTS AND AMINO ACID SYSTEMS

It is now well accepted that a whole range of excitatory amino acid antagonists have anticonvulsant properties and some of these may provide a basis for developing new drugs for the treatment of epilepsy, but what of the drugs currently in use as effective clinical anticonvulsants? Are these drugs also interacting with amino acid neurotransmitter systems in a way which would support a role for excitatory amino acids in seizure episodes? Chronic treatment of rats with several typical anticonvulsants failed to produce any consistent effect on brain levels of excitatory amino acids.[125] Some of these drugs did, however, inhibit the release of ^{3}H-D-ASP from cortex slices, a more meaningful measure of neurotransmitter function.[36,125] Further study of the phenytoin inhibition of amino acid release indicated that the effect was due to actions mediated by Na^+ or Ca^{2+} ions.[43] The anticonvulsant action of sodium valproate has been most often linked to enhancement of GABA levels. When the effects of valproate on ^{14}C-glucose incorporation into amino acids were examined, this anticonvulsant not only produced the expected increase in GABA, but also decreased glutamate and aspartate levels or formation in some brain regions.[81,82] These experimental results suggest that an action, particularly on aspartatergic neurotransmission, could equally explain the anticonvulsant action of this drug.[83]

The anticonvulsant effectiveness of A1 receptor antagonists and the observed depression of aspartate metabolism by some anticonvulsants have led some researchers to suggest that abnormalities in aspartate-mediated transmission is perhaps widely involved in epileptiform activity.[84] However, binding studies do not support the suggestion that aspartate is the only endogenous ligand of A1 receptor sites.[85,86]

When seizures were induced by glutamate injected into the lateral ventricle, a range of anticonvulsants, namely, phenytoin, phenobarbitone, chlorodiazepoxide, and valproic acid, all blocked the convulsive activity.[59] When seizures were induced by the specific A1 agonist NMDA, valproate and diazepam were effectively anticonvulsant, but phenytoin and phenobarbitone did not block the convulsions.[87] Simple excitation of hippocampal cells by superfusion of hippocampal slices with glutamate was not blocked by APV, even at concentrations seven times those which blocked NMDA responses in the same study. These results suggest that glutamate-induced excitation and seizures arising in the hippocampus may primarily involve mechanisms distinct from those mediated by A1 receptors. Recently a new anticonvulsant drug, Lamotrigine, which is chemically unrelated to those in current use, has been through preclinical trials and preliminary studies in healthy volunteers and patients with epilepsy.[135] The results are very encouraging. It was revealed to be one of the most potent anticonvulsants yet tested with a profile similar to that of phenytoin, but with a longer duration of action and less likely to impair motor performance. All tests so far indicate that only toxicity is very low or absent. Neurochemical studies suggest that Lamotrigine acts at voltage-sensitive sodium channels to stabilize neuronal membranes and inhibit neurotransmitter release, principally glutamate.

XI. OTHER EXCITANT AMINO ACIDS AND PEPTIDES

A number of other endogenous agonists for amino acid receptors have been detected in mammalian brains which could mediate the abnormal excitation of seizures. These include L-cysteine sulfinate and peptides of dicarboxylic acids such as *N*-acetyl-Asp-Glu or β-hydroxybutyl-Asp-Asp-Glu. However, the same limitation raised for quinolinic acid must apply to most of these substances. Namely, because of their low concentration (the *N*-acetyldipeptide excepted),[127] a means by which they can rapidly induce excitation of a sufficiently large group of key neurons must be established. For this reason, glutamate and aspartate remain the most likely candidates for excitatory neurotransmitter involvement in seizure activity. To support this point, a number of convulsants known to act primarily by nonexcitatory amino acid mechanisms do release glutamate and/or aspartate during seizure. These include penicillin and pentylenetetrazole,[118] methoxypridoxine,[119] 3-mercaptopropinic acid,[36] and bicuculline[120] (see above). Similarly, pyridoxal phosphate involvement in seizure has generally been thought to be through its role in GABA metabolism, but recently it has been shown that the excitatory amino acids could be potentiated at the same time[102] and that these latter changes could explain the seizure susceptibility of DBA/2 mice.[103]

XII. KINDLED EPILEPSY AND EXCITATORY AMINO ACIDS

Kindling is an animal model which is increasingly used for studying the biochemistry and pharmacology of epilepsy because it offers some unique features relative to other alternatives. The process of kindling seizures and the physiological consequences have been the subject of several recent reviews.[104-106] Briefly, kindling involves the application of low-level electrical or chemical stimulation of a limited brain region repeatedly at long intervals (hours or days). The seizures develop gradually over many stimulations (10 to 40), beginning with limbic behavioral signs (excessive jaw movements and grooming, attentional changes, etc.) and progressing to generalized tonic-clonic convulsions of differing severity. The progressive development of the kindled epilepsy can be monitored by recording the length, amplitude, and pattern of the electrical afterdischarges through electrodes positioned at the kindling site (Figure 2). The limbic brain regions, hippocampus, and amygdala or their afferent inputs are the most easily kindled and hence are most frequently used for these studies. The seizures are limited in duration to 1 to 2 min, and once kindled, the animals can undergo a long sequence of seizures on different occasions without ill effects, so each animal can be used as its own control in pharmacological studies. The kindling develops slowly, which provides the opportunity to study the neurochemical changes associated with the onset of the seizures and development of the electrical afterdischarge. The most important advantage provided by this model is that the kindled effect is long-lasting so that any alterations in brain function associated with kindling can be separated from the immediate consequences of the seizures.

In rodent models, the seizures are not spontaneous and can be precipitated at will by applying the electrical or chemical stimulus.

Neurochemical changes in many systems in the brain including the amino acids have been reported to occur during the kindling process. Direct measures of levels of the neurotransmitter amino acids 1 week after the last seizure with electrical[107] or chemical[108] kindling showed no changes in glutamate, aspartate, or any other amino acid. A more recent study showed that glutamine, the glutamate precursor, did decrease in the brains of rat kindled by stimulation of the frontal cortex.[109] This could reflect an enhanced release of glutamate in the fully kindled animal similar to that reported for kindled amygdala.[41] Further progress in determining whether there is any presynaptic effect of kindling on amino acid neurotransmitter function will require the measurement of release at the site of stimulation throughout the kindling process. However, an enhancement of veratrine-stimulated glutamate release by isolated, kindled, focal cortex has been reported.[136]

Most studies of amino acid function in kindling have concentrated on postsynaptic receptor changes. An increase in the number of A2 excitatory amino acid receptor sites after kindling by stimulation of the angular bundle afferents to the hippocampus has been reported.[110] However, a more recent study[111] failed to show any change in ^{3}H-glutamate binding or its presynaptic uptake after kindling of entorhinal cortex. This may reflect the different nature of the kindling sites involved, but it is important to consider such findings in light of the relative contribution of each receptor subtype to the total glutamate binding in each region. Kainic acid binding in the hippocampal CA3 region was shown to be dramatically reduced 1 day after the last kindled seizure induced using two different stimulation sites.[112] These short-term changes in receptor populations may implicate the excitatory amino acids in the development of kindling, but such changes must be shown to be long term if they are to account for the permanent changes which are unique to this model of epilepsy.

Direct evidence for amino neurotransmitter involvement in kindling is limited, but there are good reasons to continue these studies. Recent electrophysiological evidence indicates the primary epileptiform activity in the kindled hippocampus is not due to diminished inhibitory responses as is the case with many other models of epilepsy.[113] This suggests that the primary changes initiating the abnormal excitation at the kindled site involve an exaggerated response to the conditioning excitatory input. This is supported by the fact that afterdischarges associated with kindled seizures and recorded at the site of stimulation were completely blocked by excitatory amino acid antagonists even though the drugs were administered via the lateral ventricle and therefore some distance from this site[41,42] (Figure 2). Evidence that these antagonists can block the primary neuronal events initiating epileptiform activity has also been shown in spinal preparations. Here, paroxysmal ventral root depolarizations (the equivalent of PDS in the hippocampus) were totally prevented by a series of amino acid antagonists.[114] It is unlikely to be through coincidence that the phenomenon called long-term potentiation (LTP), which is thought to be the physiological mechanism of learning in the hippocampus, is closely linked with excitatory amino acid pathways.[115] LTP is the enduring increase in the efficacy of synaptic conduction in hippocampal pathways which follows application of a brief high-frequency burst of impulses. It is likely to involve similar categories of synaptic modification to those underlying the changes occurring during the kindling process. The implication of excitatory amino acid pathways comes from the observed sustained increases in glutamate release which follow LTP[115] and the blockade of LTP induction by the adenosine agonist 2-chloroadenosine.[116] This agent not only markedly reduced perforant path evoked potentials, it also blocked glutamate release. Also, A1 receptor antagonists selectively block the long-lasting changes in responsiveness (LTP) without affecting the normal excitation.[117] These findings strongly implicate a presynaptic contribution in the mechanisms leading to LTP. The overall similarities between LTP and kindling suggest that they may have synaptic mechanisms in common, and glutamate may be the key transmitter involved in both cases.

XIII. CORRELATION AND CONCLUSION

An important connection between kindling events, excitotoxic lesions, and epilepsy may be revealed if glutamate itself could be shown to induce chemical kindling when applied locally briefly in small amounts at subthreshold levels over a period of hours or days. Exaggerated release of this endogenous excitant would in this way be able to mediate both acute excitation and chronic epilepsy through the formation of epileptogenic lesions.

The electrophysiology of epileptic foci, including kindled foci, often suggests that the epileptic neuron is not intrinsically hyperexcitable, suggesting that the properties of the neuronal membrane itself are not altered.[128] These neurons commonly display a characteristic large depolarization shift called the PDS.[113,128,129] These recurrent shifts of depolarization take the neuronal membrane through the critical firing level, causing increased spike discharge and often a subsequent gross depolarization with inactivity due to cathodal block. This is followed by a period of hyperpolarization before onset of the next PDS. Their cause is unknown, but evidence points to their being synaptically generated and graded in nature, appearing to be greatly enhanced excitatory postsynaptic potentials. Such enhanced excitatory activity could result from endogenous agonists, such as glutamate, rapidly accumulating in the interstitial fluid.[5]

ADDENDUM

Recently MK-801 (Merk Sharp and Dohme), a potent anticonvulsant drug which also has anxiolytic and sympathomimetic properties, has been shown to act at functionally distinct sites in the mammalian brain which are related to the NMDA subtype of excitatory amino acid receptor. This suggests that its anticonvulsant properties are related to its interaction at NMDA receptors and adds further evidence implicating these receptors in the basic mechanisms generating and limiting the spread of epileptic seizures.[137] In this respect, the discovery of a new, selective, and even more potent NMDA antagonist, CPP (3-[(±)-2-carboxypiperazin-4-yl] propyl-1-phosphonic acid), may provide us with a basis for new and more effective anticonvulsant drugs.[138]

A recent study[141] has demonstrated that CPP, APH, and GAMS (in order of potency) are able to prevent the development of the kindling response when microinjected into the kindling site 15 min before delivery of the conditioning electrical stimulus. Both the afterdischarges and the associated behavioral changes were inhibited. These observations show that excitatory amino acid receptors are involved in the fundamental neural processes establishing the kindled focus as well as the hyperactivity expressed in the mature epileptic condition, (i.e., they are antiepileptogenic). These two kinds of action by antagonists may be due both to blockade of postsynaptic excitatory amino acid receptors and/or inhibitory action at presynaptic receptors which normally enhance synaptic release of glutamate (Figure 3), i.e., positive feedback receptors. Evidence for this latter action of the antagonists in hippocampus has recently been published.[19]

REFERENCES

1. **Hayashi, T.**, The effects of sodium glutamate on the nervous system, *Keio J. Med.*, 3, 183, 1954.
2. **Monaghan, D. T. and Cotman, C. W.**, Anatomical organisation of NMDA, kainate and quisqualate receptors, in *Excitatory Amino Acids*, Roberts, P. J., Storm-Mathisen, J., and Bradford, H. F., Eds., Macmillan, London, 1986.
3. **Fonnum, F.**, Glutamate: a neurotransmitter in mammalian brain, *J. Neurochem.*, 42, 1, 1984.
4. **Cotman, C. W., Foster, A. C., and Lanthorn, T.**, An overview of glutamate as a neurotransmitter, in *Glutamate as a Neurotransmitter*, Di Chiara, G. and Gessa, G. B., Eds., Raven Press, New York, 1981, chap. 1.
5. **Bradford, H. F. and Dodd, P. R.**, Biochemistry and basic mechanisms in epilepsy, in *Biochemistry and Neurological Disease*, Davison, A. N., Ed., Blackwell Scientific, Oxford, 1976, chap. 3.
6. **Fonnum, F., Soreide, A., Kvale, I., Walker, J., and Walaas, I.**, Glutamate in cortical fibres, in *Glutamate as a Neurotransmitter*, Di Chiara, G. and Gessa, G. B., Eds., Raven Press, New York, 1981, chap. 2.
7. **Fagg, G. E. and Foster, A. C.**, Amino acid neurotransmitters and their pathways in the mammalian CNS, *Neuroscience*, 9, 701, 1983.
8. **Hennecke, A. and Weichert, G.**, Seizures and the dose of L-glutamic acid in rats, *Epilepsia*, 11, 327, 1970.
9. **Johnston, G. A. R.**, Convulsions induced in 10-day-old rats by intraperitoneal injections of monosodium glutamate and related excitant amino acids, *Biochem. Pharmacol.*, 22, 137, 1973.
10. **Bhagaran, H. N., Coursin, D. B., and Stewart, C. W.**, Monosodium glutamate induces convulsive disorders in the rat, *Nature (London)*, 232, 275, 1971.
11. **Bradford, H. F. and Dodd, P. R.**, Convulsive effects of amino acids and folic acid, *Biochem. Pharmacol.*, 26, 253, 1976.
12. **Olney, J. W. and Price, M. T.**, Excitotoxic amino acids as neuroendocrine probes, in *Kainic Acid as a Tool in Neurobiology*, McGeer, E. G., Olney, J. W., and McGeer, P. L., Eds., Raven Press, New York, 1978, chap. 13.
13. **Schwarcz, R., Collins, J. F., and Parks, D. A.**, α-Amino-ω-phosphono-carboxylates block ibotenate but not kainate neurotoxicity in rat hippocampus, *Neurosci. Lett.*, 33, 85, 1982.
14. **Olney, J. W.**, Neurotoxicity of excitatory amino acids, in *Kainic Acid as a Tool in Neurobiology*, McGeer, E. G., Olney, J. W., and McGeer, P. L., Eds., Raven Press, New York, 1978, chap. 5.
15. **Benveniste, H., Drejer, J., Schousboe, A., and Diemer, N. H.**, Elevation of the extracellular concentrations of glutamate and aspartate in rat hippocampus during transient cerebral ischemia monitored by intracerebral microdialysis, *J. Neurochem.*, 43, 1369, 1984.
16. **Rothman, S.**, Synaptic release of excitatory amino acid neurotransmitters mediates anoxic neuronal death, *J. Neurosci.*, 4, 1884, 1984.
17. **Simon, R. P., Swain, J. H., Griffiths, T., and Meldrum, B. S.**, Blockade of N-methyl-D-aspartate receptors may protect against ischemic damage in the brain, *Science*, 226, 850, 1984.
18. **Berdichevsky, E., Riveros, N., Sanchez-Armass, S., and Orrego, F.**, Kainate, N-methylaspartate and other excitatory amino acids increase calcium influx into rat brain cortex cells *in vitro*, *Neurosci. Lett.*, 36, 75, 1983.
19. **Crowder, J., Croucher, M., Bradford, H. F., and Collins, J. F.**, NMDA antagonists block ^{45}Ca-uptake and amino acid release induced by depolarizing agents and excitatory amino acids, *J. Neurochem.*, 48, 1917, 1987.
20. **Van Harreveld, A. and Fifkova, E.**, Mechanisms involved in spreading depression, *J. Neurochem.*, 18, 2145, 1973.
21. **Phillis, J. W. and Ochs, S.**, Excitation and depression of cortical neurones during spreading depression, *Exp. Brain Res.*, 12, 132, 1971.
22. **Koyama, I.**, Amino acids in the cobalt-induced epileptogenic and nonepileptogenic cat's cortex, *Can. J. Physiol. Pharmacol.*, 50, 740, 1972.
23. **van Gelder, N. M. and Courtois, A.**, Close correlation between changing content of specific amino acids in epileptogenic cortex of cats and severity of epilepsy, *Brain Res.*, 43, 477, 1972.
24. **van Gelder, N. M.**, Antagonism by taurine of cobalt-induced epilepsy in cat and mouse, *Brain Res.*, 47, 157, 1972.
25. **Dodd, P. R. and Bradford, H. F.**, Release of amino acids from the maturing cobalt-epileptic focus, *Brain Res.*, 111, 377, 1976.
26. **van Gelder, N. M., Siatitsas, I., Menini, C., and Gloor, P.**, Feline generalised penicillin epilepsy: changes of glutamic and taurine parallel the progressive increase in excitability of the cortex, *Epilepsia*, 24, 200, 1983.
27. **Gottesfeld. Z. and Elazar, Z.**, GABA and glutamate in different EEG stages of the penicillin focus, *Nature (London)*, 240, 478, 1972.

28. **Koyama, I.,** Amino acid concentration in the brain of a cat with spontaneous chronic epilepsy, *J. Tokyo Wom. Med. Coll.*, 48, 428, 1978.
29. **Koyama, I. and Jasper, H.,** Amino acid content of chronic undercut cortex of the cat in relation to electrical afterdischarge: comparison with cobalt epileptogenic lesions, *Can. J. Physiol. Pharmacol.*, 55, 523, 1977.
30. **Dunn, A. and Guiddita, A.,** Long term effect of electroshock on the metabolism of glucose in the mouse brain, *Brain Res.*, 27, 418, 1971.
31. **Yoshino, Y. and Elliot, K. A. C.,** Incorporation of carbon atoms from glucose into free amino acids in the brain under normal and altered conditions, *Can. J. Physiol. Pharmacol.*, 48, 228, 1970.
32. **van Gelder, M. M., Sherwin, A. L., and Rasmussen, T.,** Amino acid content of epileptogenic human brain: focal versus surrounding regions, *Brain Res.*, 40, 385, 1972.
33. **van Gelder, N. M., Sherwin, H. L., Sacks, C., and Andermann, F.,** Biochemical observations following administration of taurine to patients with epilepsy, *Brain Res.*, 94, 297, 1975.
34. **Perry, T. L. and Hansen, S.,** Amino acid abnormalities in epileptogenic foci, *Neurology*, 31, 872, 1981.
35. **van Gelder, N. M.,** Glutamic acid in chronically hyperirritable nervous tissue, in *Advances in Epileptology: 13th Epilepsy International Symposium*, Akimoto, H., Hazamatsuri, H., Seino, M., and Wand, A., Eds., Raven Press, New York, 1981, 323.
36. **Skerritt, J. H. and Johnson, G. A. R.,** Inhibition of amino acid transmitter release from rat brain slices by phenytoin and related anticonvulsants, *Clin. Exp. Physiol. Pharmacol.*, 10, 527, 1983.
37. **Baba, A., Okumura, S., Mizo, H., and Iwata, H.,** Inhibition by diazepam and GABA and depolarization-induced release of [^{14}C]-cysteine sulphinate and [^{3}H]-glutamate in rat hippocampal slices, *J. Neurochem.*, 40, 280, 1983.
38. **Coutinho-Netto, J., Abdul-Ghani, A. S., Collins, J. F., and Bradford, H. F.,** Is glutamate a trigger factor in epileptic hyperactivity?, *Epilepsia*, 22, 289, 1981.
39. **Croucher, M. J., Collins, J. F., and Meldrum, B. S.,** Anticonvulsant action of excitatory amino acid antagonists, *Science*, 216, 899, 1982.
40. **Meldrum, B. S., Croucher, M. J., Badman, G., and Collins, J. F.,** Antiepileptic action of excitatory amino acid antagonists in the photosensitive baboon, *Papio papio, Neurosci. Lett.*, 39, 101, 1983.
41. **Peterson, D. W., Collins, J. F., and Bradford, H. F.,** The kindled amygdala model of epilepsy: anticonvulsant action of amino acid antagonists, *Brain Res.*, 275, 169, 1983.
42. **Peterson, D. W., Collins, J. F., and Bradford, H. F.,** Anticonvulsant action of amino acid antagonists against kindled hippocampal seizures, *Brain Res.*, 311, 176, 1984.
43. **Crowder, J. and Bradford, H. F.,** Common anticonvulsants inhibit ^{45}Ca-uptake and transmitter amino acid release, *Epilepsia*, 28, 378, 1987.
44. **Drejer, J., Benveniste, H., Diemer, N. H., and Schousboe, A.,** Cellular origin of ischemia-induced glutamate release from brain tissue *in vivo* and *in vitro, J. Neurochem.*, 45, 145, 1985.
45. **Peterson, D. W., Collins, J. F., and Bradford, H. F.,** Transmitter amino acids and their antagonists in epilepsy, in *Glutamine, Glutamate and GABA in the Central Nervous System*, Hertz, L., Kvamme, E., McGeer, E. G., and Schousboe, A., Eds., Alan R. Liss, New York, 1983, 643.
46. **Lloyd, K. G., Munari, C., Bossi, L., and Morselli, P. L.,** GABA hypothesis of human epilepsy: neurochemical evidence from surgically resected identified foci, in *Neurotransmitters, Seizures and Epilepsy II*, Fariello, R. G., Morselli, P. L., Lloyd, K. G., Quesney, L. F., and Engel, J., Eds., Raven Press, New York, 1984, 285.
47. **Meldrum, B. S.,** Epilepsy, in *The Molecular Basis of Neuropathology*, Davison, A. N. and Thompson, R. H. S., Edward Arnold, London, 1981, chap. 11.
48. **Davidson, N.,** *Neurotransmitter Amino Acids*, Academic Press, London, 1976.
49. **Pisa, M., Sandberg, P. R., Corcoran, M. E., and Fibiger, H. C.,** Spontaneous recurrent seizures after intracerebral injections of kainic acid in rat: a possible model of human temporal lobe epilepsy, *Brain Res.*, 200, 481, 1980.
50. **Bradford, H. F. and McIlwain, H.,** Ionic basis for the depolarization of cerebral tissues by excitatory amino acids, *J. Neurochem.*, 13, 1163, 1966.
51. **Farber, L.,** The role of calcium in cell death, *Life Sci.*, 29, 1289, 1981.
52. **Cabral, R. J., King, T. T., and Scott, D. F.,** Epilepsy after two different neurosurgical approaches to the treatment of ruptured intracranial aneurysm, *J. Neurol. Neurosurg. Psychiatr.*, 39, 1052, 1979.
53. **North, J. B., Penhall, R. K., Hanieh, A., Hann, C. S., Challen, R. G,. and Frewin, D. B.,** Postoperative epilepsy: a double-blind trial of phenytoin after craniotomy, *Lancet*, 1, 384, 1980.
54. **McBean, G. J. and Roberts, P. J.,** Neurotoxicity of L-glutamate and DL-threo-3-hydroxyaspartate in rat striatum, *J. Neurochem.*, 44, 247, 1985.
55. **Kohler, C., Schwarcz, R., and Fuxe, K.,** Perforant path transactions protect hippocampal granule cells from kainate lesions, *Neurosci. Lett.*, 10, 241, 1978.
56. **Biziere, K. and Coyle, J. T.,** Effects of cortical ablation on the neurotoxicity and receptor binding of kainic acid in striatum, *J. Neurosci. Res.*, 4, 383, 1979.

57. **Ferkany, J. W., Zaczak, R., and Coyle, J. T.**, Kainic acid stimulates excitatory amino acid neurotransmitter release at presynaptic terminals, *Nature (London)*, 298, 757, 1982.
58. **Dodd, P. R., Bradford, H. F., Abdul-Ghani, A. S., Cox, D. W. G., and Coutinho-Netto, J.**, Release of amino acids from chronic epileptic and sub-epileptic foci *in vivo*, *Brain Res.*, 193, 505, 1980.
59. **Stone, W. E. and Javid, M. J.**, Effects of anticonvulsants and other agents on seizures induced by intracerebral L-glutamate, *Brain Res.*, 264, 165, 1983.
60. **Morrell, F.**, Secondary epileptogenic lesions, *Epilepsia*, 1, 538, 1960.
61. **Watkins, J. C. and Evans, R. H.**, Excitatory amino acid transmitters, *Annu. Rev. Pharmacol. Toxicol.*, 21, 165, 1981.
62. **Zaczek, R. and Coyle, J. T.**, Excitatory amino acid analogues: neurotoxicity and seizures, *Neuropharmacology*, 21, 15, 1982.
63. **Colton, C. A. and Colton, J. S.**, Blockade of hyperbaric oxygen induced seizures by excitatory amino acid antagonists, *Can. J. Physiol. Pharmacol.*, 63, 519, 1985.
64. **Croucher, M. J., Meldrum, B. S., and Collins, J. F.**, Anticonvulsant and proconvulsant properties of a series of structural isomers of piperidine dicarboxylic acid, *Neuropharmacology*, 23, 467, 1984.
65. **Jones, A. W., Croucher, M. J., Meldrum, B. S., and Watkins, J. C.**, Suppression of audiogenic seizures in DBA/2 mice by two new dipeptides NMDA receptor antagonists, *Neurosci. Lett.*, 45, 157, 1984.
66. **Koerner, J. F. and Cotman, C. W.**, Micromolar L-2-amino-4-phosphonobutyric acid selectively inhibits perforant path synapses from lateral entorhinal cortex, *Brain Res.*, 216, 192, 1981.
67. **Butcher, S. P., Collins, J. F., and Roberts, P. J.**, Characterization of the binding of DL-(^{3}H)-2-amino-4-phosphonobutyrate to L-glutamate-sensitive sites on rat brain synaptic membranes, *Br. J. Pharmacol.*, 80, 355, 1983.
68. **Fagg, G. E., Foster, A. C., Mena, E. E., and Cotman, C. W.**, Chloride and calcium reveal a pharmacologically distinct population of L-glutamate binding sites in synaptic membranes: correspondence between biochemical and electrophysiological data, *J. Neurosci.*, 2, 958, 1982.
69. **Monaghan, D. T., Holets, V. R., Toy, D. W., and Cotman, C. W.**, Anatomical distribution of four pharmacologically distinct ^{3}H-L-glutamate binding sites, *Nature (London)*, 306, 176, 1983.
70. **Robinson, M. B., Crooks, S. L., Johnson, R. L., and Koerner, J. F.**, Displacement of DL-(^{3}H)-2-amino-4-phosphonobutanoic acid binding with methyl-substituted APB analogues and glutamate agonists, *Biochemistry*, 24, 2401, 1985.
71. **Pin, J. P., Bockaert, J., and Recasesn, M.**, The Ca^{2+}/Cl^- dependent L-(^{3}H)glutamate binding: a new receptor or a particular transport process?, *FEBS Lett.*, 175, 31, 1984.
72. **Davis, J. and Watkins, J. C.**, Actions of D and L forms of 2-amino-5-phosphonovalerate and 2-amino-4-phosphonobutyrate in the rat spinal cord, *Brain Res.*, 235, 378, 1982.
73. **Freed, W. J.**, Selective inhibition of homocysteine-induced seizures by glutamic acid diethyl ester and other glutamate esters, *Epilepsia*, 26, 30, 1985.
74. **Collins, J. F., Dixon, A. J., Badman, G., De Sarro, G., Chapman, A. G., Hart, G. P., and Meldrum, B. S.**, Kainic acid derivatives with anticonvulsant activity, *Neurosci. Lett.*, 51, 371, 1984.
75. **Turski, L., Meldrum, B. S., and Collins, J. F.**, Anticonvulsant action of β-kainic acid in mice: is β-kainic acid a N-methyl-D-aspartate antagonists?, *Brain Res.*, 336, 162, 1985.
76. **Stone, T. W. and Collins, J. F.**, β-Kainic acid is not an amino acid antagonist, *J. Pharm. Pharmacol.*, 37, 668, 1985.
77. **Croucher, M. J., Meldrum, B. S., Jones, A. W., and Watkins, J. C.**, γ-D-Glutamylamino-methylsulfonic acid (GAMS), a kainate and quisqualate antagonist prevents sound-induced seizures in DBA/2 mice, *Brain Res.*, 322, 111, 1984.
78. **Erez, U., Frank, H., Goldberg, O., Cohen, A., and Teichberg, V. I.**, Anticonvulsant properties of 3-hydroxy-2-quinoxalinecarboxylic acid, a newly formed antagonist of excitatory amino acids, *Eur. J. Pharmacol.*, 110, 31, 1985.
79. **Patsalos, P. N. and Lasielles, P. T.**, Change in regional brain levels of amino acid putative neurotransmitters after prolonged treatment with the anticonvulsant drugs diphenylhydantoin, phenobarbitone, sodium valproate, ethosuximide and sulthiame in the rat, *J. Neurochem.*, 36, 688, 1981.
80. **Colasanti, B. K. and Craig, C. R.**, Brain concentration and synthesis rate of biogenic amines during chronic cobalt experimental epilepsy in the rat, *Neuropharmacology*, 12, 221, 1973.
81. **Schechter, P. J., Tranier, Y., and Grove, J.**, Effects of n-dipropylacetate on amino acid concentrations in mouse brain: correlation with anticonvulsant activity, *J. Neurochem.*, 31, 1325, 1978.
82. **Chapman, A. G., Riley, K., Evans, M. C., and Meldrum, B. S.**, Acute effects of sodium valproate and γ-vinyl GABA on regional amino acid metabolism in the rat brain: incorporation of 2(^{14}C) glucose into amino acids, *Neurochem. Res.*, 7, 1089, 1982.
83. **Chapman, A., Keane, P. E., Meldrum, B. S., Simiand, J., and Verniures, J. C.**, Mechanism of anticonvulsant action of valproate, *Prog. Neurobiol.*, 19, 315, 1982.

84. **Meldrum, B. S. and Chapman, A. G.,** Excitatory amino acids and anticonvulsant drug action, in *Glutamine, Glutamate, and GABA in the Central Nervous System,* Hertz, L., Kvamme, E., McGeer, E. G., and Schousboe, A., Eds., Alan R. Liss, New York, 1983, 625.
85. **Ferkany, J. W. and Coyle, J. T.,** Specific binding of (^{3}H) + -2-amino-7-phosphonoheptanoic acid to rat brain membranes *in vitro, Life Sci.,* 33, 1295, 1983.
86. **Olverman, H. J., Jones, A. W., and Watkins, J. C.,** L-Glutamate has a higher affinity than other amino acids for (^{3}H)-D-AP5 binding sites in rat brain membranes, *Nature (London),* 307, 460, 1984.
87. **Czuczwar, S. J., Frey, H. H., and Loscher, W.,** Antagonism of N-methyl-D,L aspartatic acid induced convulsions by antiepileptic drugs and other agents, *Eur. J. Pharmacol.,* 108, 273, 1985.
88. **Nadler, J. V.,** Kainic acid as a tool for the study of temporal lobe epilepsy, *Life Sci.,* 29, 2031, 1981.
89. **Ben-Ari, Y., Tremblay, E., Ottersen, O. P., and Meldrum, B. S.,** The role of epileptic activity in hippocampal and "remote" cerebral lesions induced by kainic acid, *Brain Res.,* 191, 79, 1980.
90. **Sperk, G., Lassman, H., Baran, H., Seitelberger, F., and Hornykeiwicz, O.,** Kainic acid induced seizures: dose-relationship of behavioural, neurochemical and histopathological changes, *Brain Res.,* 338, 289, 1985.
91. **Cavalheiro, E. A., Calderazzo Filko, L. L., Riche, D., Feldrum, S., and Le Gal la Salle, G.,** Amygdaloid lesions increase the toxicity of intrahippocampal kainic acid injections and reduce the late occurrence of spontaneous recurrent seizures in rats, *Brain Res.,* 262, 201, 1983.
92. **Westbrook, G. L. and Lothman, E. W.,** Cellular and synaptic basis of kainic acid-induced hippocampal epileptiform activity, *Brain Res.,* 273, 97, 1983.
93. **Lothman, E. R. and Collins, R. C.,** Kainic acid induced limbic seizures: metabolic, behavioural, electroencephalographic and neuropathological correlates, *Brain Res.,* 218, 299, 1981.
94. **Engel, J., Jr.,** Epileptic brain damage: how much excitement can a limbic neuron take?, *Trends Neurosci.,* 6, 356, 1983.
95. **Sawada, S. and Yamamoto, C.,** Fast and slow depolarizing potentials induced by short pulses of kainic acid in hippocampal neurons, *Brain Res.,* 324, 279, 1984.
96. **Ferkany, J. W. and Coyle, J. T.,** Kainic acid selectivity stimulates the release of endogenous excitatory acidic amino acids, *J. Pharmacol. Exp. Ther.,* 225, 399, 1983.
97. **Franck, J. E. and Schwartzkroin, P. A.,** Do kainate-lesioned hippocampi become epileptogenic?, *Brain Res.,* 329, 309, 1985.
98. **Stone, T. W. and Perkins, M. N.,** Quinolinic acid: a potent endogenous excitant at amino acid receptors in CNS, *Eur. J. Pharmacol.,* 72, 711, 1981.
99. **Lapin, I. P.,** Convulsant action of intracerebroventricularly administered 1-kynurinine sulfate, quinolinic acid and other derivatives of succinic acid, and effects of amino acids: structure activity relationships, *Neuropharmacology,* 21, 1227, 1982.
100. **Perkins, M. N. and Stone, T. W.,** On the interaction of 2-amino-7-phosphono-heptanoic acid and quinolinic acid in mice, *Eur. J. Pharmacol.,* 89, 297, 1983.
101. **Stone, T. W. and Connick, J. H.,** Quinolinic acid and other kynurenines in the central nervous system, *Neuroscience,* 15, 597, 1985.
102. **Norris, D. J., Murphy, R. A., and Chung, S. H.,** Alteration of amino acid metabolism in epileptogenic mice by elevation of brain pyridoxal phosphate, *J. Neurochem.,* 44, 1403, 1985.
103. **Chung, S. H. and Cox, R. A.,** Determination of pyridoxal phosphate levels in the brains of audiogenic and normal mice, *Neurochem. Res.,* 8, 1245, 1983.
104. **Wada, J. A., Eds.,** *Kindling Two,* Raven Press, New York, 1981.
105. **Goddard, G. V.,** The kindling model of epilepsy, *Trends Neurosci.,* 6, 275, 1983.
106. **McNamara, J. O.,** Kindling: an animal model of complex partial epilepsy, *Ann. Neurol. Suppl.,* 16, S72, 1984.
107. **Fabisiak, J. P. and Schwark, W. S.,** Cerebral free amino acids in the amygdaloid kindling model of epilepsy, *Neuropharmacology,* 21, 179, 1982.
108. **Fabisiak, J. P. and Schwark, W. S.,** Aspects of pentylenetetrazol kindling model of epileptogenesis in the rat, *Exp. Neurol.,* 78, 7, 1982.
109. **Miller, A. A., O'Donnell, R. A., and Webster, R. A.,** Reduced cortical glutamine concentrations in electrically kindled rats, *J. Neurochem.,* 41, 1492, 1983.
110. **Savage, D. D., Werling, L. L., Nadler, J. V., and McNamara, J. O.,** Selective increase in L-(^{3}H)glutamate binding to a quisqualate-sensitive site on hippocampal synaptic membranes after angular bundle kindling, *Eur. J. Pharmacol.,* 85, 255, 1982.
111. **Selvin, J. T. and Ferrara, L. P.,** Lack of effect of entorhinal kindling on L-(^{3}H)-glutamic acid presynaptic uptake and postsynaptic binding in hippocampus, *Exp. Neurol.,* 89, 48, 1985.
112. **Savage, D. D., Nadler, J. V., and McNamara, J. O.,** Reduced kainic binding in rat hippocampal formation after limbic kindling, *Brain Res.,* 323, 128, 1984.
113. **Oliver, M. W. and Miller, J. J.,** Inhibitory processes of hippocampal CA1 pyramidal neurons following kindling-induced epilepsy in the rat, *Can. J. Physiol. Pharmacol.,* 63, 872, 1985.

114. **Ryan, G. P., Hackman, J. C., and Davidoff, R. A.,** Spinal seizures and excitatory amino acid-mediated synaptic transmission, *Neurosci. Lett.*, 44, 161, 1984.
115. **Dolphin, A. C., Errington, M. L., and Bliss, T. V.,** Long term potentiation of the perforant path *in vivo* is associated with increased glutamate release, *Nature (London)*, 297, 496, 1982.
116. **Dolphin, A. C.,** The adenosine agonist 2-chloroadenosine inhibits the induction of long term potentiation of the perforant path, *Neurosci. Lett.*, 39, 83, 1983.
117. **Collingridge, G. L., Kehl, S. J., and McLennan, H.,** Excitatory amino acids in synaptic transmission in the Schaffer collateral-commissural pathway of the rat hippocampus, *J. Physiol.*, 334, 33, 1983.
118. **De Boer, Th., Stoof, C., and van Duijn, H.,** The effects of convulsant and anticonvulsant drugs on the release of radiolabeled GABA, glutamate, noradrenaline, serotonin and acetylcholine from rat cortical slices, *Brain Res.*, 253, 153, 1982.
119. **Nitsch, C., Schmude, B., and Haug, P.,** Alterations in the content of amino acid neurotransmitters before the onset and during the course of methoxypyridoxine-induced seizures in individual rabbit brain regions, *J. Neurochem.*, 40, 1571, 1983.
120. **Chapman, A. G. and Evans, M. C.,** Cortical GABA turnover during bicuculline seizures in rats, *J. Neurochem.*, 41, 886, 1983.
121. **Bradford, H. F., Young, A. M. J., and Crowder, J. M.,** Continuous glutamate leakage from brain cells is balanced by compensatory high-affinity reuptake transport, *Neurosci. Lett.*, 81, 296, 1987.
122. **Czuazwar, S. J. and Meldrum, B. S.,** Protection against chemical induced seizures by 2-amino-7-phosphonoheptanoic acid, *Eur. J. Pharmacol.*, 83, 335, 1982.
123. **Fagg, G. E.,** L-Glutamate, excitatory amino acid receptors and brain function, *Trends Neurosci.*, 8, 207, 1985.
124. **Turski, L., Meldrum, B. S., Jones, A. W., and Watkins, J. C.,** Anticonvulsant action of stereoisomers of γ-glutamylaminomethylsulfonic acid in mice, *Eur. J. Pharmacol.*, 111, 279, 1985.
125. **Skerrit, J. H. and Johnston, G. A. R.,** Enhancement of excitant amino acid release from rat brain slices by the convulsant 3-mercaptoproprionic acid, *Brain Res.*, 258, 165, 1983.
126. **Young, A. M. J. and Bradford, H. F.,** Excitatory amino acid neurotransmitters in the corticostriate pathway. Studies using intracerebral microdialysis *in vivo*, *J. Neurochem.*, in press.
127. **Zaczek, R., Koller, K., Cotter, R., Heller, D., and Coyle, J. T.,** N-acetylaspartylglutamate: an endogenous peptide with high affinity for a brain "glutamate" receptor, *Proc. Natl. Acad. Sci. U.S.A.*, 80, 116, 1983.
128. **Ajmone-Marsan, C.,** Acute effects of topical epileptogenic agents, in *Basic Mechanisms in the Epilepsies*, Jasper, H. H., Ward, A. A., and Pope, A., Eds., Churchill Livingstone, London, 1969, chap. 11.
129. **Johnston, D.,** The synaptic nature of paroxysomal depolarization shifts in hippocampal neurons, *Ann.*
130. **Olney, J. W., Price, T. M., Samson, L., and Labruyere, J.,** The role of specific ions in glutamate neurotoxicity, *Neurosci. Lett.*, 65, 65, 1986.
131. **Nicklas, W. J., Zeevalk, G. D., and Hyndman, A. G.,** The role of the Cl^- in mechanisms of action of excitotoxins, in *Proc. Am. Soc. Neurochem.*, Abstr. 251, Montreal, 1986, 218.
132. **Choi, D. W.,** Glutamate neurotoxicity in cortical cell culture is calcium-dependent, *Neurosci. Lett.*, 58, 293, 1985.
133. **Garthwaithe, G. and Gaithwaithe, J.,** Neurotoxicology of excitatory amino acid receptor agonists in rat cerebellar slices: dependence on calcium concentrations, *Neurosci. Lett.*, 66, 193, 1986.
134. **Rothman, S. M. and Olney, J. W.,** Glutamate and the pathophysiology of hypoxic ischemic brain damage, *Ann. Neurol.*, 19, 105, 1986.
135. **Miller, A. A., Sawyer, D. A., Roth, B., Peck, A. W., Leach, M. J., Wheatley, P. L., Parsons, D. N., and Morgan, R. J. I.,** Lamotrigine, in *New Anticonvulsant Drugs*, Meldrum, B. S. and Porter, R. J., Eds., John Libbey, London, 1986, 165.
136. **Leach, M. J., Martin, E. M., Miller, A. A., and O'Donnell, R. A.,** Changes in cortical amino acids during electrical kindling, *Neuropharmacology*, 24, 937, 1986.
137. **Wong, E. H. F., Kemp, J. A., Prestley, T., Knight, A. R., Woodruff, G. N., and Iversen, L. L.,** The anticonvulsant MK-801 is a potent N-methyl-D-aspartate antagonist, *Proc. Natl. Acad. Sci. U.S.A.*, 83, 7104, 1986.
138. **Davies, J., Evans, R. H., Herrling, P. L., Jones, A. W., Olverman, H. J., Pook, P., and Watkins, J. C.,** CPP, a new potent and selective NMDA antagonist, *Brain Res.*, 382, 169, 1986.
139. **Geddes, J. W., Chang-Chui, H., Cooper, S. M., Lott, I. T., and Cotman, C. W.,** Density and distribution of NMDA receptors in the human hippocampus in Alzheimer's disease, *Brain Res.*, 399, 156, 1986.
140. **Monaghan, D. T. and Cotman, C. W.,** Identification and properties of NMDA-receptors in rat brain synaptic plasma membranes, *Proc. Natl. Acad. Sci. U.S.A.*, 83, 7532, 1986.
141. **Croucher, M. J. and Bradford, H. F.,** *Eur. J. Pharmacol.*, in press; *Brit. J. Pharmacol.*, 97, 865P, 1987; *Biochem. Soc. Trans.*, 14, 910, 1987.

Chapter 24

EXPERIMENTAL AND CLINICAL STATES OF HYPERAMMONIA: ALTERATIONS IN GLUTAMATE AND GLUTAMINE

Oliver W. Cass and Leslie Zieve

TABLE OF CONTENTS

I. DISEASES ASSOCIATED WITH HYPERAMMONIA

Much of the ammonia produced in the body is formed in the large intestine from the action of bacteria on the contents of the colon[1] and in the small intestine from the oxidation of glutamine by the epithelium.[2] As a result, the ammonia concentration in portal blood is normally several times that of venous or arterial blood. A healthy liver converts most of the ammonia to urea through the Krebs-Henseleit cycle. Small amounts of ammonia are normally released from the kidney, muscle, and brain.[3,4]

Many conditions, outlined in Table 1, can result in ammonia excess. Much arises from an inability to dispose of ammonia, either because of specific enzyme defects (as in the congenital urea cycle abnormalities) or hepatic dysfunction (as in hepatic failure from viral hepatitis). Some is caused by a shunt of portal blood around the liver to the systemic circulation (as in cirrhosis or in surgically created portacaval shunts). Little is ordinarily due to an increased ammonia load (as in excessive protein ingestion). Most hyperammonia is due to a combination of all four problems. Patients with cirrhosis and hepatic dysfunction have a decreased ability to make urea, develop shunts between the portal and systemic circulations, and occasionally digest large amounts of ammonia-producing blood when they bleed from varices, gastritis, or peptic ulcers.

Experimental studies have made use of several methods for simulating hyperammonia states. Injections of ammonium salts provide an increase in the ammonia load. Portasystemic shunts have simulated bypass of the liver, and hepatectomy or hepatic necrosis by various toxins and manipulations have simulated clinical hepatic failure. In addition, experimental cirrhosis and fulminant hepatic failure have been produced. The changes in various tissues accompanying hyperammonia states produced by each of the above models will be considered in the sections that follow, along with the changes observed in human diseases.

II. HYPERAMMONIA IN VARIOUS TISSUES: EFFECTS ON GLUTAMATE AND GLUTAMINE

A. Blood

In normal rats, a dose less than the LD_{50} of an ammonium salt was rapidly cleared from the blood so that the ammonia levels returned to normal in 5 to 10 min.[5] A small rise in plasma glutamine occurred. Dogs made comatose with ammonium salts had a fall in most plasma amino acids, except glutamine, which increased slightly.[6] By permitting portal blood rich in ammonia to enter the systemic circulation, portacaval shunts increased the concentration of ammonia in the arterial blood.[7-12] Plasma glutamine rose and glutamate decreased or remained unchanged. Tyrosine, phenylalanine, and free tryptophan (aromatic amino acids, AAA) increased in blood, whereas isoleucine, leucine, and valine (branched-chain amino acids, BCAAs) decreased.[6,10,11,13,14] In animals with portacaval shunts, the acute administration of ammonium salts, meat, urease, or urea elevated the blood ammonia further.[6-8,12,15] Plasma glutamine tended to increase slightly, but glutamate remained normal.[7] When animals were subjected to total hepatectomy, urea production fell and the blood ammonia concentration rose.[16-18] Blood glutamine increased, but glutamic acid remained unchanged.[10,17-19] Many other amino acids rose as well. The AAAs and methionine were the most elevated, while BCAAs were normal or slightly decreased.[10,18,20]

Humans with acute fulminant hepatic failure showed a rise in both plasma glutamine and glutamic acid.[21-23] In addition, most plasma amino acids were elevated, the highest being the AAAs and methionine. Presumably, the necrosed liver tissue was the main source of these amino acids. The BCAAs were normal or slightly decreased. Patients with cirrhosis without overt encephalopathy had a slightly increased blood ammonia. One half of the patients showed a rise in both glutamine and glutamic acid.[23,24] The AAAs and methionine were slightly elevated, and the BCAAs were slightly depressed.[24,25]

Table 1
SYNDROMES OF HYPERAMMONIA[116-118]

- Inherited disorders
  - Urea cycle enzyme deficiencies
    - Carbamyl phosphate synthetase deficiency
    - Ornithine transcarbamylase deficiency
    - Argininosuccinate synthetase deficiency
    - Argininosuccinase deficiency
    - Arginase deficiency
  - Aminoacidopathies
    - Ornithinemia
    - Periodic hyperlysinemia
    - Lysinuric protein intolerance
  - Organic acidemias
    - Propionic acidemia
    - Methylmalonic acidemia
    - Isovaleric acidemia
    - Multiple carboxylase deficiency
    - Glutaric acidemia II
    - Ethylmalonic acidemia
  - Other enzyme deficiencies
    - Hereditary protein intolerance
    - Carnitine deficiency
- Noninherited disorders
  - Neonatal hyperammonemia
  - Reye's syndrome
  - Reye's-like syndromes due to
    - Pyrrolizidine alkaloids
    - Jamaica vomiting sickness
    - Diethyltoluamide
    - Aflatoxin
    - Valproic acid
  - Hepatic insufficiency
    - Fulminant hepatic failure
    - Other acute liver disease
    - Chronic liver disease
    - Surgical removal of liver
  - Portacaval shunts
  - Exogenous ammonium salts
  - Hyperalimentation
  - Infection with urea-splitting organism
  - Ureterosigmoidostomy
  - Muscular contraction
    - Severe exercise
    - Seizures
    - Mild exercise in cirrhotics

The blood ammonia in patients with cirrhosis and overt encephalopathy increased with the grade of encephalopathy.[26-28] However, there was enough overlap among grades so the blood ammonia could not be used to diagnose encephalopathy or distinguish severity. There was no clear relation between plasma glutamine and severity of encephalopathy in patients with decompensated cirrhosis.[29] However, glutamic acid, aspartic acid, methionine, phenylalanine, and tyrosine were usually significantly elevated, while the BCAAs were reduced.[22,25,29,30]

To summarize, all of the models of hyperammonemia result in a rise in plasma glutamine, with the exception of chronic liver disease. Glutamate is elevated only in fulminant hepatic failure and the most severe grades of encephalopathy. The AAAs tend to be elevated and the BCAAs tend to be decreased in all of the models that have been discussed. The ratio, AAA/BCAA, is a fair indicator of severity of hepatic failure, but is not an indicator of the presence or severity of hepatic encephalopathy.[31]

B. Cerebrospinal Fluid

Cerebrospinal fluid (CSF) is more difficult to obtain than blood, but it reflects the metabolism of the brain and the changes in metabolism induced by liver failure better. When dogs were given sufficient ammonium salt to induce coma, the CSF ammonia and glutamine rose while glutamate remained normal. All other amino acids, including the BCAAs, decreased or were unchanged.[6] Following portacaval shunt in rats and dogs, CSF ammonia and glutamine increased while glutamate remained unchanged.[6,8,32] AAAs increased substantially and BCAAs increased slightly. When shunted animals were given ammonium salts, CSF ammonia increased further, glutamine stayed at its postshunt level, and glutamate decreased.

Humans with fulminant hepatic failure had marked elevations of both glutamine and glutamate.[21,23] All other amino acids except aspartate and arginine were elevated as well. In chronic liver disease without overt encephalopathy, CSF ammonia and glutamine were elevated, while glutamic acid was normal.[23,33] In encephalopathic patients, CSF ammonia and glutamine were more strikingly elevated.[23,29,33,34] The level of CSF glutamine correlated well with the severity of encephalopathy as determined by clinical examination[34,35] or electroencephalography.[36] Furthermore, following clinical improvement in encephalopathy during treatment with BCAA mixtures, the CSF glutamine improved.[29] There was a poor correlation between the CSF glutamine level and the blood ammonia.[33]

C. Brain

1. Normal Ammonia Metabolism[37]

Ammonia is taken up from the blood and CSF and is also generated by the brain.[4,38,39] The fractional uptake of ammonia from the blood was constant at about 25% over a 1000-fold concentration range, but increased with higher pH.[38,40] The ammonium ion is thus less diffusible than the gas. When tracer doses of ^{13}N ammonia were infused into humans, they entered the brain rapidly at a time when most of the label was still in ammonia.[4] Thus, the tracer entered as ammonia, not as a metabolic product. Very little of the label left the brain as the specific activity in plasma declined, implying that the label was trapped in the brain in a less readily diffusible form.

Arteriovenous concentration differences across the cerebral circulation indicated that the brain normally released glutamine into the circulation.[41] Ammonia and glutamate were the source of the glutamine. When rats received intracarotid ^{13}N ammonia and were killed by freezing their brains within 5 sec of the injection, 57% of the label was recovered in glutamine.[38] This rapid trapping mechanism can be overwhelmed by chronic exposure to an elevated blood ammonia, resulting in an increase in the brain ammonia concentration.[8] The trapping mechanism appeared limited to a specific, small compartment in the brain. When coma-inducing doses of ^{15}N ammonia were infused into the carotid arteries of cats, the highest specific activity of ^{15}N was found in the glutamine amide N.[42,43] In addition, the glutamine α amino N had a higher specific activity than the α amino N of glutamate. This implied that the ammonia N entered a small, metabolically active pool of glutamate which did not exchange readily with most of the brain glutamate.[44] Tracer studies of ^{13}N-labeled ammonia confirmed these results in normal rats.[38]

When brain glutamine synthetase was inhibited by pretreatment with methionine sulfoximine, more ^{13}N ammonia was recovered unmetabolized, and the specific activity of glutamate was higher than that of glutamine.[38] Inhibition of glutamine synthetase abolished the distinct metabolic pools and restored the precursor-product relationship of glutamate and glutamine expected elsewhere in the body. The small compartment of metabolically active glutamate was probably the astrocytes. They are suitably located between the blood vessels of the brain and neurons, and glutamine synthetase can be localized to them by immunohistochemical techniques using light and electron microscopy.[45,46] Conditions resulting in long-term exposure of the brain to hyperammonia alone produced typical pathologic changes in the astrocytes like those seen in hepatic coma.[47-50]

2. *Neuronal Effects of Ammonia*

Ammonia is a neurotoxin. Ammonium ions can substitute for Na^+ or K^+, initially producing neuronal stimulation, but eventually depolarization block, by increasing the egress of K^+ from neurons and altering Cl^- permeability.[51,52] Na^+, K^+-ATPase was stimulated by ammonium ion, and acute hyperammonemia induced an increase of the enzyme in mouse brains.[53] Ammonia inactivated the extrusion of Cl^- from neurons and thus abolished the suppression of action potential generation resulting from postsynaptic inhibition.[52,54-56] It inhibited glutaminase in synaptosomes and thus reduced glutamate and the precursor pools for GABA and aspartate.[51,57,58] Ammonia may also alter the stores of glutamate in synaptosomes. Slices of rat hippocampus released glutamate when incubated with ammonia and were subsequently unable to release normal amounts of glutamate when exposed to K^+.[59]

3. *Effect of Acute Ammonia Intoxication*

Ammonium salts given to normal animals produced hyperventilation, respiratory alkalosis, convulsions, stupor, and coma.[60,61] Carbon dioxide inhalation was protective. Brain oxidative metabolism was stimulated without a change in energy balance.[62-65] The utilization of glucose was increased, but the cerebral blood flow and oxygen consumption were variably affected. Glycolytic intermediates including pyruvate and lactate increased; α-ketoglutarate (2-oxoglutarate) was reduced. Glutamine and AAAs increased, but glutamate decreased or remained unchanged.[66-68] Brain content of dopamine and norepinephrine was reduced, and brain content of serotonin and 5HIAA was increased. Convulsions due to NH_4^+ were associated with at least a tenfold increase in brain ammonia.[60] Convulsions from other agents were associated with a modest rise in brain ammonia. Paradoxically, in the presence of acute centrilobular liver damage less cerebral glutamine was formed in response to an NH_4^+ load.[69] Mammals exposed to hyperammonia for variable time periods showed proliferation of Alzheimer type II astrocytes similar to that seen in hepatic coma.[47-50]

4. *Effect of Portacaval Shunts and of Shunts Plus an Ammonium Load*

As with blood ammonia, brain ammonia and glutamine increased while glutamate remained normal or slightly depressed.[7,8,32,70] Tryptophan and tyrosine also increased in the brain, but aspartate, asparagine, GABA, and alanine remained unchanged.[8,13] Several weeks after shunts in rats, cerebral blood flow increased, but the cerebral metabolic rate for oxygen was unaffected.[7] Normal animals developed increased Alzheimer type II astrocytes in their cerebrum after the shunt.[15]

The acute administration of ammonium salts to rats and dogs with portasystemic shunts caused lethargy, coma, and EEG changes similar to those of hepatic coma.[7,8,12] In shunted dogs and humans, the same occurred with a large protein load.[12,15,71] Similar results followed urease injection or the ingestion of urea in dogs.[12] In rats with shunts, the additional ammonium load elevated the brain ammonia and glutamine further.[7,8] Brain aspartate and glutamate decreased; alanine and asparagine increased; and GABA remained unchanged.[8] When rats with portacaval anastomoses were given the additional ammonia, cerebral blood flow and oxygen utilization declined, the lactate/pyruvate concentration ratio in the brain increased, and the $NAD^+/NADH$ concentration ratio and ATP concentration declined.[7,8] These abnormalities may all be due to a failure to oxidize cytoplasmic NADH. NADH cannot diffuse into mitochondria, so reducing equivalents must be transported by the malate/aspartate shuttle. If glutamate is converted to glutamine during ammonia metabolism, it may not be available to exchange for aspartate in the shuttle. Indeed, isolated brain mitochondria could not function without all the intermediates of this shuttle being present in the culture medium.[72]

5. Effect of Hepatic Failure

Dogs and rats subjected to total hepatectomy became lethargic 12 to 15 hr after the operation and progressed to coma, usually within 24 hr.[16,17,73] Brain ammonia and glutamine increased, while glutamate and aspartate remained normal or decreased.[10,20,74,75] Most other amino acids also increased, especially the AAAs. Similar results were seen after massive hepatic ischemia.[76,77] In addition to glutamate and aspartate, excitatory neurotransmitters such as dopamine and norepinephrine were reduced, whereas GABA and cyclic AMP were unchanged or slightly increased.[10,20,76]

Patients dying with fulminant hepatic failure had marked elevations of most amino acids in their brains, but especially methionine and the BCAAs, which were increased over 16 times normal.[21] Even glutamic acid increased 1.5-fold. Patients with chronic liver failure due to cirrhosis severe enough to develop episodes of encephalopathy had increased brain ammonia utilization rates compared to normals or those with mild cirrhosis.[4] They trapped more of administered $^{13}NH_3$, and the utilization rate was linearly related to the arterial ammonia concentration. When such patients died, they had elevated brain ammonia and glutamine and, less prominently, histidine and isoleucine, but normal brain glutamate.[78]

Rabbits with galactosamine-induced hepatic coma had decreased numbers of excitatory neurotransmitter receptors on brain synaptic membranes, including those for glutamic acid, aspartic acid, and kainic acid, and increased numbers of inhibitory neurotransmitter receptors, including those for glycine and GABA.[79-81] Similar increases in GABA receptors occurred in rats with galactosamine-induced hepatic encephalopathy, dogs with portasystemic shunts and encephalopathy, and patients who died with hepatic encephalopathy.[82-84] A synthetic "hepatic" coma induced in rats by the synergistic action on the neuron of the three main toxins (NH_4^+, methanethiol, and octanoic acid) that accumulate during human or experimental hepatic failure was also associated with increased brain GABA receptors.[85] It appears that neuronal damage by the hepatic failure toxins decreases the susceptibility of the neuron to glutamate and other excitatory neurotransmitters and increases its susceptibility to inhibitory neurotransmitters like GABA and tranquilizers like benzodiazepines.

The GABA receptor on postsynaptic neurons is composed of a supramolecular complex with binding sites for three classes of compounds: GABA, benzodiazepines, and barbiturates.[86,87] Humans with hepatic encephalopathy are notoriously sensitive to both of the latter two classes of drugs. Rabbits and rats with galactosamine-induced hepatic coma had increased numbers of receptors for flunitrazepam and diazepam, respectively.[81,88] It is therefore of interest that the administration of a benzodiazepine antagonist (CGS 8216, Ciba-Geigy) to rats with galactosamine-induced hepatic encephalopathy resulted in a return to normal alertness and activity, normal visual evoked potential, and a reduction of the number and affinity of benzodiazepine receptors.[88] Also another shorter-acting benzodiazepine antagonist (RO 15-1788, Hoffman-LaRoche) transiently reversed coma in the first three patients with hepatic coma that received trial injections of the compound.[89]

D. Liver

Ammonia is an important constituent in the synthesis of urea, glutamine, aspartate, or glutamate by the liver. Most of the ammonia presented to the cat liver was metabolized to urea.[42] The next largest amount appeared in aspartic acid, then in glutamic acid, and least in glutamine. Little is known of the ability of the liver in normal humans to regulate glutamine and glutamate levels in response to hyperammonemia. Although the liver contains glutamate dehydrogenase and glutamine synthetase, glutamine synthesis was not stimulated by ammonium chloride in the isolated, perfused rat liver.[90,91] Instead, the ammonia was converted to urea and only small amounts of glutamine and glutamate were formed. When the intracellular concentrations of glutamate and ammonia were increased by perfusion with histidine, glutamine synthesis remained normal. The addition of NH_4Cl to a culture of rat hepatocytes increased glucose synthesis and the intracellular concentration of glucose and glutamate, while α-ketoglutarate decreased.[92,93] Unlike the brain, a normal liver subjected to hyperammonemia responds mainly with urea and glucose, not glutamine synthesis.

Most states of hyperammonemia are due either to specific defects in the urea cycle or to severe liver disease, in which all liver functions, including urea synthesis, are decreased. Indeed, hyperammonemia and encephalopathy in cirrhosis were inversely correlated with the maximal rate of urea synthesis.[94] There is some evidence that liver ammonia metabolism may be altered by portacaval shunts. Glutamine synthetase activity declined by two thirds 3 to 4 weeks after rats had an end-to-side portacaval shunt.[95] The small fraction of ammonia converted to glutamine in the liver may thus be even less after this procedure. In hepatic failure, the pattern of hepatocyte damage may determine the disposition of ammonia in the liver. Glutamine synthetase was shown to be localized in the area surrounding the central vein in the rat.[96] When pericentral hepatocyte damage was produced in rats with carbon tetrachloride, urea synthesis remained normal, while glutamine production fell.[97]

E. Muscle

Studies of arteriovenous differences across resting normal skeletal muscle showed that ammonia was nearly constant but glutamate and glutamine were taken up.[98-100] When the rat hindquarter or diaphragm was exposed to an ammonia load, the tissue content of glutamate fell and the release of glutamine increased by as much as 80%.[101,102] In addition, the conversion of ^{14}C-leucine to carbon dioxide was increased. This may account for the decrease in plasma BCAAs seen in several hyperammonemic states discussed above. Following hepatectomy, the concentration of glutamine in dog skeletal muscle rose and that of glutamate fell.[74] In humans with cirrhosis, forearm muscle extracted ammonia, and the rate of extraction increased with increasing muscle mass.[98] In patients with cirrhosis and coma due to an increased ammonia load from GI bleeding, the extraction rate was proportional to the blood ammonia concentration.[100] When the liver is damaged so that it cannot remove its normal share of ammonia from the portal circulation, the muscle may assume the increased burden of removing ammonia. Unfortunately, most of the chronic liver diseases have associated muscle wasting, which only compounds the hyperammonemia. If nutrition and muscle mass could be preserved in chronic liver disease, one might anticipate less hyperammonemia and encephalopathy.

F. Intestine

The intestine accounts for much of the ammonia produced in the body. About one half comes from bacterial metabolism of nitrogenous compounds in the colon, and one half comes from the metabolism of glutamine by small intestinal epithelium.[103] In certain hyperammonemic states, the intestine continues to produce ammonia. Dogs and rats subjected to total hepatectomy showed a progressive rise in blood ammonia that could be prevented by evisceration.[104,105] Some of this was undoubtedly produced in the colon, as neomycin, a nonabsorbable antibiotic, can improve hepatic encephalopathy and lower blood ammonia when given to cirrhotics.[106,107] However, germ-free dogs with portacaval shunts, and hepatectomized germ-free rats had similar elevations in blood ammonia, indicating that small intestinal ammonia production must be important.[108,110] The relative contributions of the colon and small intestine to ammonia production in hyperammonia are uncertain. Whether the increased plasma glutamine seen in many hyperammonia states causes an increase in small intestinal production of ammonia has not been studied.

G. Kidney and Other Tissues

The kidney produces ammonia from glutamine and transports it into the renal tubule by several processes.[3] This production was increased in acidosis, as the kidney used ammonium ions to excrete an acid load.[111] About one half of the ammonia produced in normal people was excreted in the urine, but the other half was released into the circulation.[112] This contributed to the blood ammonia. Although the kidney has been thoroughly studied in acidosis, the effect of hyperammonemia on its ammonia genesis and glutamine metabolism is not clear.

Other tissues, including bone,[113] fibroblasts,[114] and lymphocytes,[115] consume glutamine and produce ammonia, but their contribution to ammonia production is minimal and their response to hyperammonemia is unstudied.

III. SUMMARY

Almost all states of hyperammonia resulted in tissue excess of glutamine while glutamate remained unchanged or depressed. With massive liver necrosis both glutamine and glutamate were elevated. AAAs were usually increased when glutamine increased, and BCAAs were decreased. A good correlation was seen between the increase in brain and CSF glutamine and the presence and severity of encephalopathy in hepatic failure. Neuronal deterioration due to the excess of ammonia and other hepatic failure toxins, such as mercaptans, fatty acids, and phenols, resulted in reduced responsiveness to glutamate and other excitatory neurotransmitters and increased susceptibility to inhibitory neurotransmitters such as GABA and depressive agents such as the benzodiazepines.

REFERENCES

1. **Wolpert, E., Phillips, M. D., Summerskill, W. H. J.,** Ammonia production in the human colon. Effects of cleansing, neomycin and acetohydroxamic acid, *N. Engl. J. Med.*, 283, 159, 1970.
2. **Souba, W. W., Smith, R. J., and Wilmore, D. W.,** Glutamine metabolism by the intestinal tract, *J. Par Ent Nutr.*, 9, 608, 1985.
3. **Good, D. W. and Knepper, M. A.,** Ammonia transport in the mammalian kidney, *Am. J. Physiol.*, 248, F459, 1985.
4. **Lockwood, A. H., McDonald, J. M., Rieman, R. E., Golbard, A. S., Laughlin, J. S., Duffy, T. F., and Plum, F.,** The dynamics of ammonia metabolism in man, *J. Clin. Invest.*, 63, 449, 1979.
5. **Rosado, A., Flores, G., Mora, J., and Soberon, J. G.,** Distribution of an ammonia load in the normal rat, *Am. J. Physiol.*, 203, 37, 1962.
6. **Schafer, K., Ukida, M., Steffen, C., Bode, C., Zelder, D., and Bode, J. C.,** Effect of ammonia on plasma and cerebrospinal fluid amino acids in dogs with and without portacaval anastomoses, *Res. Exp. Med.*, 185, 35, 1985.
7. **Gjedde, A., Lockwood, A. H., Duffy, T. E., and Plum, F.,** Cerebral blood flow and metabolism in chronically hyperammonemic rats: effect of an acute ammonia challenge, *Ann. Neurol.*, 3, 325, 1978.
8. **Hindfelt, B., Plum, F., and Duffy, T. E.,** Effect of acute ammonia intoxication on cerebral metabolism in rats with portacaval shunts, *J. Clin. Invest.*, 59, 386, 1977.
9. **Kline, D. G., Crook, J. N., and Nance, F. C.,** Eck fistula encephalopathy. Long term studies in primates, *Ann. Surg.*, 173, 97, 1971.
10. **Mans, A. M., Saunders, S. J., Kirsch, R. E., and Biebuyck, J. F.,** Correlation of plasma and brain amino acid and putative neurotransmitter alterations during acute hepatic coma in the rat, *J. Neurochem.*, 32, 285, 1979.
11. **Mans, A. M., Biebuyck, J. F., Davis, D. W., and Hawkins, R. A.,** Portacaval anastamosis: brain and plasma metabolite abnormalities and the effect of nutritional therapy, *J. Neurochem.*, 43, 697, 1984.

12. **Riddell, A. G., Kopple, P. N., and McDermott, W. V.,** The etiology of "meat intoxication" in the Eck fistula dog, *Surgery,* 36, 675, 1954.
13. **James, J. H., Hodgman, J. M., Funovics, J. M., Voschimura, N., and Fischer, J. E.,** Brain tryptophan, plasma free tryptophan and distribution of plasma neutral amino acids, *Metabolism,* 25, 471, 1976.
14. **Smith, A. R., Rossi-Fanelli, F., Ziparo, V., James, J. H., Perelle, B. A., and Fischer, J. E.,** Alterations in plasma and CSF amino acids and metabolites in hepatic coma, *Ann.Surg.,* 187, 343, 1978.
15. **Taylor, P., Schoene, W. C., Reid, W. A., Jr., and von Lichtenberg, F.,** Quantitative changes in astrocytes after portacaval shunting, *Arch. Pathol. Lab. Med.,* 103, 82, 1979.
16. **Bollman, J. L., Mann, F. C., and Magath, T. B.,** Effect of total removal of the liver on the formation of urea, *Am. J. Physiol.,* 69, 371, 1924.
17. **Drapanas, T., McMenamy, R. H., Adler, W. J., and Vang, J. O.,** Intermediary metabolism following hepatectomy in dogs, *Ann. Surg.,* 162, 621, 1965.
18. **McMenamy, R. H., Vang, J., and Drapanas, T.,** Amino acid and alpha keto acid concentrations in plasma and blood of the liverless dog, *Am. J. Physiol.,* 209, 1046, 1965.
19. **Flock, E. V., Block, M. A., Mann, F. C., Grindley, J. H., and Bollman, J. L.,** The effect of glucose on the amino acids of plasma after total hepatectomy, *J. Biol. Chem.,* 198, 427, 1952.
20. **Holmin, T., Agardh, C. D., Alinder, G., Werlin, P., and Heltberg, B.,** The influence of total hepatectomy on cerebral energy state, ammonia-related amino acids of the brain and plasma amino acids in the rat, *Eur. J. Clin. Invest.,* 13, 215, 1983.
21. **Record, C. O., Buxton, B., Chase, R. A., Curzon, G., Murray-Lyon, I. M., and Williams, R.,** Plasma and brain amino acids in fulminant hepatic failure and their relationship to hepatic encephalopathy, *Eur. J. Clin. Invest.,* 6, 387, 1976.
22. **Rosen, Hm., Yoshimura, N., Hodgman, J. M., and Fischer, J. E.,** Plasma amino acid patterns in hepatic encephalopathy of differing etiology, *Gastroenterology,* 72, 483, 1977.
23. **Watanabe, A., Takei, N., Higahi, T., Nakatsukasa, H., Fujinara, M., Sakala, T., and Nagashima, H.,** Glutamic acid and glutamine levels in serum and cerebrospinal fluid in hepatic encephalopathy, *Biochem. Med.,* 32, 225, 1984.
24. **Ansley, J. D., Isaacs, J. W., Rikkers, L. F., Kutner, M. D., Nordlinger, B. M., and Rudman, D.,** Quantitative tests of nitrogen metabolism in cirrhosis: relation to other manifestations of liver disease, *Gastroenterology,* 75, 570, 1978.
25. **Cascino, A., Cangiano, C., Calcaterra, V., Rossi-Fanelli, F., and Capocaccia, L.,** Plasma amino acid imbalance in patients with liver disease, *Dig. Dis.,* 23, 591, 1978.
26. **Seegmiller, J. E., Schwartz, R., and Davidson, C.,** The plasma "ammonia" and glutamine content in patients with hepatic coma, *J. Clin. Invest.,* 33, 984, 1954.
27. **Stahl, J.,** Studies of the blood ammonia in liver disease: its diagnostic, prognostic, and therapeutic significance, *Ann. Intern. Med.,* 58, 1, 1963.
28. **Tyor, M. P. and Sieker, H. O.,** Biochemical, blood gas and peripheral circulatory alterations in hepatic coma, *Am. J. Med.,* 59, 50, 1959.
29. **Rossle, M., Luft, M., Herz, R., Klein, D., Lehmann, M., and Gerok, W.,** Amino acid, ammonia and neurotransmitter concentrations in hepatic encephalopathy: serial analysis in plasma and cerebrospinal fluid during treatment with an adapted amino acid solution, *Klin. Wochenschr.,* 62, 867, 1984.
30. **Fischer, J. E., Yoshimura, N., Aguirre, A., James, J. H., Cummings, M. G., Abel, R. M., and Diendoerger, F.,** Plasma amino acids in patients with hepatic encephalopathy. Effects of amino acid infusions, *Am. J. Surg.,* 127, 40, 1974.
31. **Morgan, M. Y., Milsom, J. P., and Sherlock, S.,** Plasma ratio of valine, leucine, and isoleucine to phenylalanine and tyrosine in liver disease, *Gut,* 19, 1068, 1978.
32. **Bollman, J. L., Flock, E. V., Grindley, J. H., Bickford, R. G., and Lichtenheld, F. R.,** Coma with increased amino acids of brain and cerebrospinal fluid in dogs with Eck's fistula, *Arch. Surg.,* 75, 405, 1957.
33. **Caesar, J.,** Levels of glutamine and ammonia and the pH of cerebrospinal fluid and plasma in patients with liver disease, *Clin. Sci.,* 22, 33, 1962.
34. **Hourani, B. T., Hamlin, E. M., and Reynolds, T. B.,** Cerebrospinal fluid glutamine as a measure of hepatic encephalopathy, *Arch. Intern. Med.,* 127, 1033, 1971.
35. **Velasco, M., Vaccarezza, A., Galvez, S., Chioffalo, N., Klinger, J., Katz, R., and Cuevas, N.,** Amonia sanguineo, glutamina en el liquido cefaloraquideo, electroen-cefalograma y examen neurologico en relacion a los estados de la conciencia en los enfermos hepaticos, *Rev. Med. Chile,* 89, 767, 1961.
36. **Oei, L. T., Kuys, J., Lombarts, A. J. P. F., Goor, C., and Endtz, L. J.,** Cerebrospinal fluid glutamine levels and eeg findings in patients with hepatic encephalopathy, *Clin. Neurol. Neurosurg.,* 81, 59, 1979.
37. **Duffy, T. E., Plum, F., and Cooper, A. J. L.,** Cerebral ammonia metabolism in vivo, in *Metabolic Relationships Between Glutamine, Glutamate and GABA in the CNS,* Hertz, L., Kvamme, E., McGeer, E. G., and Schousboe, E., Eds., Alan R. Liss, New York, 1983, 371.

38. **Cooper, A. J. L., McDonald, J. M., Gelbard, A. S., Gledhill, R. F., and Duffy, T. E.,** The metabolic fate of ^{13}N-labelled ammonia in rat brain, *J. Biol. Chem.*, 254, 4982, 1979.
39. **Weil-Malherbe, H. and Gordon, J.,** Amino acid metabolism and ammonia formation in brain slices, *J. Neurochem.*, 18, 1659, 1971.
40. **Raichle, M. E. and Larson, K. B.,** The significance of the NH_3-NH_4^+ equilibrium on the passage of ^{13}N ammonia from blood to brain, *Circ. Res.*, 48, 913, 1981.
41. **Hills, A. G., Reid, E. L., and Kerr, W. D.,** Circulatory transport of L-glutamine in fasted mammals: cellular sources of urine ammonia, *Am. J. Physiol.*, 223, 1470, 1967.
42. **Berl, S., Takagaki, G., Clarke, D. D., and Waelsch, H.,** Metabolic compartments in vivo. Ammonia and glutamic acid metabolism in brain and liver, *J. Biol. Chem.*, 237, 2562, 1962.
43. **Takagaki, G., Berl, S., Clarke, D. D., Purpura, D. P., and Waelsch, H.,** Glutamic acid metabolism in brain and liver during infusion with ammonia labelled with nitrogen-15, *Nature (London)*, 189, 326, 1961.
44. **Berl, S. and Clarke, D. D.,** Compartmentation of amino acid metabolism, in *Structural Neurochemistry*, Lajtha, A., Ed., Handbook of Neurochemistry Series, Vol. 2, Plenum Press, New York, 1969, 447.
45. **Norenberg, M. D. and Martinez-Hernandez, A.,** Fine structural localization of glutamine synthetase in astrocytes of rat brain, *Brain Res.*, 161, 303, 1979.
46. **Martinez-Hernandez, A., Bell, K. P., and Norenberg, M. D.,** Glutamine synthetase: glial localization in brain, *Science*, 195, 1356, 1977.
47. **Bruton, C. J., Corsellis, J. A. N., and Russell, A.,** Hereditary hyperammonemia, *Brain*, 93, 423, 1970.
48. **Cole, M., Rutherford, R. B., and Smith, F. O.,** Experimental ammonia encephalopathy in the primate, *Arch. Neurol.*, 26, 130, 1972.
49. **Gibson, G. E., Zimber, A., Krook, L., Richardson, E. P., Jr., and Viske, W. J.,** Brain histology and behavior of mice injected with urease, *J. Neuropathol. Exp. Neurol.*, 33, 201, 1974.
50. **Gutierrez, J. A. and Norenberg, M. D.,** Alzheimer II astrocytosis following methionine sulfoximine, *Arch. Neurol.*, 32, 123, 1975.
51. **Benjamin, A. M., Okamoto, K., and Quastel, J. H.,** Effects of ammonium ions on spontaneous action potentials and on contents of sodium, potassium, ammonium and chloride ions in brain in vitro, *J. Neurochem.*, 30, 131, 1978.
52. **Raabe, W. and Gumnit, R. J.,** Disinhibition in cat motocortex by ammonia, *J. Neurophysiol.*, 38, 347, 1975.
53. **Sadasivudu, B., Rao, T. I., and Murthy, C. R.,** Acute metabolic effects of ammonia in mouse brain, *Neurochem. Res.*, 2, 639, 1977.
54. **Lux, H. D.,** Ammonium and chloride extrusion: hyperpolarizing synaptic inhibition in cat spinal motoneurons, *Science*, 173, 555, 1971.
55. **Raabe, W. and Lin, S.,** Ammonia, postsynaptic inhibition and CNS-energy state, *Brain Res.*, 303, 67, 1984.
56. **Raabe, W. A. and Onstad, G. R.,** Ammonia and methionine sulfoximine intoxication, *Brain Res.*, 242, 291, 1982.
57. **Bradford, H. F. and Ward, H. K.,** Glutamine as a metabolic substrate for isolated nerve endings: inhibition by ammonium ions, *Biochem. Soc. Trans.*, 3, 1223, 1975.
58. **Bradford, H. F. and Ward, H. K.,** On glutaminase activity in mammalian synaptosomes, *Brain Res.*, 110, 115, 1976.

58a. **Benjamin, A. M.,** Control of glutaminase activity in rat brain cortex in vitro: influence of glutamate, phosphate, ammonium, calcium and hydrogen ions, *Brain Res.*, 208, 363, 1981.

59. **Hamberger, A., Jacobson, I., Molin, S. O., Nystrom, B., and Sandberg, M.,** Glutamate as a CNS neurotransmitter: the regulation of its biosynthesis by ammonium ions. A model for hepatic encephalopathy, in *Synaptic Constituents in Health and Disease*, Brzin, M., Sket, D., and Bachelard, H., Eds., Pergamon Press, Oxford, 1980, 494.
60. **Torda, C.,** Ammonium ion content and electrical activity of the brain during the preconvulsive and convulsive phases induced by various convulsants, *J. Pharmacol. Exp. Ther.*, 107, 197, 1953.
61. **Roberts, K. E., Thompson, F. G., III, Poppell, J. W., and Vanamee, P.,** Respiratory alkalosis accompanying ammonium toxicity, *J. Appl. Physiol.*, 9, 367, 1956.
62. **Clarke, G. M. and Eiseman, B.,** Studies in ammonia metabolism. IV. Biochemical changes in brain tissue of dogs during ammonia induced coma, *N. Engl. J. Med.*, 259, 178, 1958.
63. **Hawkins, R. A., Miller, A. L., Nielsen, R. C., and Veech, R. L.,** The acute action of ammonia on rat brain metabolism in vivo, *Biochem. J.*, 134, 1001, 1973.
64. **James, I. M., MacDonnell, L., and Xanalatos, C.,** Effect of ammonium salts on brain metabolism, *J. Neurol. Neurosurg. Psychiatr.*, 37, 948, 1974.
65. **McKhann, G. M. and Tower, D. B.,** Ammonia toxicity and cerebral oxidative metabolism, *Am. J. Physiol.*, 200, 420, 1961.

66. **Eiseman, B., Osofsky, H., Roberts, E., and Jelink, B.,** Cerebral free amino acids in dogs following infusion of ammonia, *J. Appl. Physiol.*, 14, 251, 1959.
67. **Tews, J. K. and Stone, W. E.,** Free amino acids and related compounds in brain and other tissues: effects of convulsant drugs, *Prog. Brain Res.*, 16, 135, 1965.
68. **Jeppsson, B., Gimmon, Z., James, J. H., Nobin, A., and Fischer, J. E.,** Infusion of NH_4^+-salts changes brain neurotransmitters, *Gastroenterology*, 79, 1028, 1980.
69. **Baraona, E., Salinas, A., Navia, E., and Orrego, H.,** Alterations of ammonia metabolism in the cerebral cortex of rats with hepatic damage induced by carbon tetrachloride, *Clin. Sci.*, 28, 201, 1965.
70. **Williams, A. H., Kyu, M. H., Fenton, J. C. B., and Cavanagh, J. B.,** The glutamate and glutamine content of rat brain after portocaval anastamosis, *J. Neurochem.*, 19, 1073, 1972.
71. **Balo, J. and Korpassy, B.,** The encephalitis of dogs with Eck fistula fed on meat, *Arch. Pathol.*, 13, 80, 1932.
72. **Dennis, S. C. and Clark, J. B.,** The regulation of glutamate metabolism by tricarboxylic acid cycle activity in rat brain mitochondria, *Biochem. J.*, 172, 155, 1978.
73. **Mann, F. C.,** The effects of complete and of partial removal of the liver, *Medicine*, 6, 419, 1927.
74. **Flock, E. V., Block, M. A., Grindley, J. H., Mann, F. C., and Bollman, J. L.,** Changes in free amino acids of brain and muscle after total hepatectomy, *J. Biol. Chem.*, 200, 529, 1953.
75. **Flock, E. V., Tyce, G. M., and Owen, C. A., Jr.,** Utilization of U-^{14}C-glucose in brain after total hepatectomy in the rat, *J. Neurochem.*, 13, 1389, 1966.
76. **Funovics, J. M. and Fischer, J. E.,** Brain energy metabolism and alterations of transmitter profiles in acute hepatic coma, *J. Neural Transm. Suppl.*, 14, 61, 1978.
77. **Mattson, W. J., Iob, V., Sloan, M., Coon, W. W., Turcotte, J. G., and Child, C. G.,** Alterations of individual free amino acids in brain during acute hepatic coma, *Surg. Gynecol. Obstet.*, 130, 263, 1970.
78. **Weiser, M., Riederer, P., and Kleinberger, G.,** Human cerebral free amino acids in hepatic coma, *J. Neural Transm. Suppl.*, 14, 95, 1978.
79. **Ferenci, P., Pappas, S. C., Munson, P. J., and Jones, E. A.,** Changes in the status of neurotransmitter receptors in a rabbit model of hepatic encephalopathy, *Hepatology*, 4, 186, 1984.
80. **Ferenci, P., Pappas, S. C., Munson, P. J., and Jones, E. A.,** Changes in glutamate receptors on synaptic membranes associated with hepatic encephalopathy or hyperammonemia in the rabbit, *Hepatology*, 4, 25, 1984.
81. **Schafer, D. F., Fowler, J. M., Munson, P. J., Thakur, A. K., Waggoner, J. G., and Jones, E. A.,** Gamma-aminobutyric acid and benzodiazepine receptors in an animal model of fulminant hepatic failure, *J. Lab. Clin. Med.*, 102, 870, 1983.
82. **Baraldi, M. and Zeneroli, M. L.,** Experimental hepatic encephalopathy: changes in the binding of gamma-aminobutyric acid, *Science*, 216, 427, 1982.
83. **Ferenci, P., Riederer, P., Jellinger, K., Schafer, D. F., and Jones, E. A.,** Changes in gamma-aminobutyric acid receptors in patients with hepatic encephalopathy, *Gastroenterology*, 84, 1371, 1983.
84. **Baraldi, M., Zeneroli, M. L., Ventura, E., Pinelli, G., Ricci, P., Santi, M., Racagni, G., Iuliano, E., Casciarri, I., Germini, M., Cavalletti, E., and Tofanetti, G.,** Portal-systemic encephalopathy in dogs: changes in brain GABA receptors and neurochemical correlates, in *Advances in Hepatic Encephalopathy and Urea Cycle Diseases*, Kleinberger, G., Ferenci, P., Riederer, P., and Thaler, H., Eds., S. Karger, Basel, 1984, 353.
85. **Baraldi, M., Pinelli, G., Ricci, P., and Zeneroli, M. L.,** Toxins in hepatic encephalopathy. The role of the synergistic effect of ammonia, mercaptans and short chain fatty acids, *Arch. Toxicol. Suppl.*, 7, 103, 1984.
86. **Guidotti, A., Toffano, G., Baraldi, M., Schwartz, J. P., and Costa, E.,** A molecular mechanism for the facilitation of GABA receptor function by benzodiazepines, in *GABA-Neurotransmitters*, Krogsgaard-Larsen, P., Scheel-Kruger, J., and Kofod, H., Eds., Academic Press, New York, 1979, 406.
87. **Guidotti, A., Baraldi, M., and Costa, E.,** 1,4-Benzodiazepines and gamma-aminobutyric acid: pharmacological and biochemical correlates, *Pharmacology*, 19, 267, 1979.
88. **Baraldi, M., Zeneroli, M. L., Ventura, E., Penne, A., Pinelli, G., Ricci, P., and Santi, M.,** Supersensitivity of benzodiazepine receptors in hepatic encephalopathy due to fulminant hepatic failure in the rat: reversal by a benzodiazepine antagonist, *Clin. Sci.*, 67, 1677, 1984.
89. Letters to the Editor. Reversal of hepatic coma by benzodiazepine antagonist (Ro 15-1788), *Lancet*, 1, 1324, 1985.
90. **Chamuleau, R. A. F. M. and Tager, J. M.,** Nitrogen metabolism in the perfused rat liver, *Biochim. Biophys. Acta*, 222, 119, 1970.
91. **Lund, P.,** Control of glutamine synthesis in rat liver, *Biochem. J.*, 124, 653, 1971.
92. **Joseph, S. K. and McGivan, J. D.,** The effect of ammonium chloride and glucagon on the metabolism of glutamine in isolated liver cells from starved rats, *Biochim. Biophys. Acta*, 543, 16, 1978.
93. **Ochs, R. S.,** Glutamine metabolism of isolated rat hepatocytes. Evidence for catecholamine activation of alpha ketoglutarate dehydrogenase, *J. Biol. Chem.*, 259, 13,004, 1984.

94. **Rudman, D., Difulco, T. J., and Galambos, J. T.,** Maximal rates of excretion and synthesis of urea in normal and cirrhotic subjects, *J. Clin. Invest.*, 52, 2241, 1973.
95. **Kennan, A. L.,** Glutamine synthetase activity in rat liver after portacaval shunt, *Endocrinology*, 74, 805, 1964.
96. **Häussinger, D.,** Hepatocyte heterogeneity in glutamine and ammonia metabolism and the role of an intercellular glutamine cycle during ureagenesis in perfused rat liver, *Eur. J. Biochem.*, 133, 269, 1983.
97. **Häussinger, D. and Gerok, W.,** Hepatocyte heterogeneity in ammonia metabolism: impairment of glutamine synthesis in CCl_4 induced liver cell necrosis with no effect on urea synthesis, *Chem. Biol. Interact.*, 48, 191, 1984.
98. **Ganda, O. P. and Ruderman, N. B.,** Muscle nitrogen metabolism in chronic hepatic insufficiency, *Metabolism*, 25, 427, 1976.
99. **Marliss, E. B., Aoki, T. T., Pozefski, T., Mast, A. S., and Cahill, G. F., Jr.,** Muscle splanchnic glutamine and glutamate metabolism in postabsorptive and starved man, *J. Clin. Invest.*, 50, 814, 1971.
100. **Webster, L. T., Jr. and Gabuzda, G. J.,** Ammonia uptake by extremities and brain in hepatic coma, *J. Clin. Invest.*, 37, 414, 1958.
101. **Chang, T. W. and Goldberg, A. L.,** The origin of alanine produced in skeletal muscle, *J. Biol. Chem.*, 253, 3677, 1978.
102. **Ruderman, N. B. and Lund, P.,** Amino acid metabolism in skeletal muscle. Regulation of glutamine and alanine release in the perfused rat hindquarter, *Isr. J. Med. Sci.*, 8, 295, 1972.
103. **Weber, F. L., Jr. and Veach, G. L.,** The importance of the small intestine in gut ammonium production in the fasting dog, *Gastroenterology*, 77, 235, 1979.
104. **Degos, F., Degos, J.-D., Bourdiau, D., Peignoux, M., Prandi, D., Roche-Sicot, J., Sicot, C., Rueff, B., and Benhamou, J. P.,** Experimental acute hepatic encephalopathy: comparison of the electroencephalographic changes in the liverless and in the eviscerated rat, *Clin. Sci. Mol. Med.*, 47, 599, 1974.
105. **Lawrence, W., Jr., Schwartz, A. E., and Randall, H. T.,** Alterations in blood ammonia in dogs following total hepatectomy and abdominal evisceration, *Surg. Gynecol. Obstet.*, 107, 69, 1958.
106. **Crossley, I. R. and Williams, R.,** Progress in the treatment of chronic portasystemic encephalopathy, *Gut*, 25, 85, 1984.
107. **Fischer, C. J. and Faloon, W. W.,** Blood ammonia levels in hepatic cirrhosis: their control by the oral administration of neomycin, *N. Engl. J. Med.*, 256, 1030, 1957.
108. **Nance, F. C. and Kline, D. G.,** Eck's fistula encephalopathy in germ-free dogs, *Ann. Surg.*, 174, 856, 1971.
109. **Nance, F. C., Batson, R. C., and Kline, D. G.,** Ammonia production in germ-free Eck fistula dogs, *Surgery*, 70, 169, 1974.
110. **Schalm, S. W. and Van der Mey, T.,** Hyperammonemic coma after hepatectomy in germ-free rats, *Gastroenterology*, 77, 231, 1983.
111. **Tannen, R. L. and Sastrasinh, S.,** Response of ammonia metabolism to acute acidosis, *Kidney Int.*, 25, 1, 1984.
112. **Reitzer, L. J., Wice, B. M., and Kennel, D.,** Evidence that glutamine, not sugar, is the major energy source for cultured HeLa cells, *J. Biol. Chem.*, 254, 2669, 1979.
113. **Biltz, R. M., Letteri, J. M., Pellegrino, E. D., Palekar, A., and Pinkus, L. M.,** Glutamine metabolism in bone, *Miner. Electrolyte Metab.*, 9, 125, 1983.
114. **Zielke, R. H., Ozand, P. T., Tildon, J. T., Sevdalian, D. A., and Cornblath, M.,** Reciprocal regulation of glucose and glutamine utilization by cultured human diploid fibroblasts, *J. Cell. Physiol.*, 95, 41, 1978.
115. **Ardawi, M. S. M. and Newsholme, E. A.,** Glutamine metabolism in lymphocytes of the rat, *Biochem. J.*, 212, 835, 1983.
116. **Flannery, D. B., Hsia, Y. E., and Wolf, B.,** Current status of hyperammonemic syndromes, *Hepatology*, 2, 495, 1982.
117. **Shih, V. E.,** Hereditary urea-cycle disorders, in *The Urea Cycle*, Grisolia, S., Baguena, R., and Mayar, F., Eds., John Wiley & Sons, New York, 1976, 367.
118. **Zieve, L.,** Hepatic encephalopathy, in *Diseases of the Liver*, 6th ed., Schiff, L. and Schiff, E. R., Eds., Lippincott, Philadelphia, in press.

Index

INDEX

A

P

Q

R

S

T

U

V

For Product Safety Concerns and Information please contact our
EU representative GPSR@taylorandfrancis.com Taylor & Francis
Verlag GmbH, Kaufingerstraße 24, 80331 München, Germany